SANITATION AMONG THE SCHEDULED TRIBES OF ARUNACHAL PRADESH

About the Authors

Dr. Bikash Bage is an Associate Professor and Head of the Department of Sociology, Rajiv Gandhi University. He is also the Nodal Officer of Tribal Research Institute, Rajiv Gandhi University.He has to his credit more than 30 research articles published in reputed and refereed Journals andin the edited books. He has presented a number of research papers in various national and international conferences and seminars in India and abroad. He has carried out more than 15Research Projects and Consultancy Projects funded by Government of India, Government of Arunachal Pradesh and other funding agencies like Indian Space and Research Organisation. He has been conferred the State Governor's award for his contribution in the field of health education.

Dr. Ashi Lama is an Associate Professor in the Department of Economics, Rajiv Gandhi University. He has to his credit more than 35 research articles published in reputed and refereed Journals andedited books. He has published one authored book, two edited volumes and two working papers. He has also presented a number of research papers in various national and international conferences. He has also carried out more than 10 Research Projects funded by Government of India, Government of Arunachal Pradesh and other funding agencies like Indian Council of Social Science Research,UGC, Centre for Development Studies, Rajiv Gandhi University and ICRIER, New Delhi.

SANITATION AMONG THE SCHEDULED TRIBES OF ARUNACHAL PRADESH

Status, Awareness Level and Hygienic Practices

Bikash Bage
Ashi Lama

Research Undertaken by
Tribal Research Institute, Rajiv Gandhi University,
Rono Hills, Doimukh

Sponsored by
Department of Social Justice Empowerment and Tribal Affairs
Government of Arunachal Pradesh, Itanagar

Funded by
Ministry of Tribal Affairs, Government of India, New Delhi

CONCEPT PUBLISHING COMPANY PVT. LTD.
NEW DELHI - 110059

ISBN: 978-93-5594-673-7

First Published 2024

Published and Printed by

Concept Publishing Company Pvt. Ltd.
A/15-16, Commercial Block, Mohan Garden
New Delhi-110059 (India)
T: +91-11-41101460
E: publishing@conceptpub.com, *W*: www.conceptpub.com

Cataloging in Publication--Courtesy: D.K. Agencies (P) Ltd. <docinfo@dkagencies.com>

Bage, Bikash, author.
Sanitation among the scheduled tribes of Arunachal Pradesh : status, awareness level and hygienic practices / Bikash Bage, Ashi Lama.
pages cm
Study conducted in Papumpare District of Arunāchal Pradesh, India.
"Tribal Research Institute, Rajiv Gandhi Univesity, Rono Hills, Doimukh ... Sponsored by Department of Social Justice Empowerment and Tribal Affairs, Government of Arunachal Pradesh, Itanagar, Funded by Ministry of Tribal Affairs, Government of India, New Delhi."
Includes bibliographical references and index.
ISBN 9789355946737

1. Papumpare (India: District)--Scheduled tribes--Health and hygiene. 2. Sanitation--India--Papumpare (District) 3. Refuse and refuse disposal--India--Papumpare (District) 4. Household surveys--India--Papumpare (District) 5. Papumpare (India : District)--Scheduled tribes 6. Attitudes. I. Lama, Ashi, author. II. Rajiv Gandhi University. Tribal Research Institute. III. Arunāchal Pradesh (India). Department of Social Justice Empowerment and Tribal Affairs, sponsoring body. IV. India. Ministry of Tribal Affairs, sponsoring body. V. Title.

LCC RA567.5.I4B34 2024 | DDC 363.7208900954163 23

Preface

Sanitation is the key to public health and quality of life of the people. It is considered to be the most basic human need for health and well-being. It encompasses public health condition related to supply of safe drinking water, access to sanitary latrines, proper drainage and sewage system as well as proper and scientific disposal of solid waste. Access to safe water and sanitation is essential not only for good health but also for alleviating poverty, safety and security of women and girl children as well as for human dignity. Access to better sanitation prevents various diseases, such as diarrhoca, typhoid, worm infections and protects human life. Proper sanitation also improves human productivity, reduces healthcare cost and promotes social and economic prosperity. Water and sanitation are related intrinsically. In view of importance of sanitation, the United Nations has declared ensuring access to safe water and sanitation as one of the 17 goals of sustainable development. Though the sanitation is very important for the well-being of people and development of a nation, a large number of people in the world, particularly in developing countries, do not have access to proper sanitation facilities. The Joint Monitoring Programme of the United Nations Children's Fund (UNICEF) and the World Health Organisation (WHO) estimated that 2 billion people in the world (26%) were without access to safe drinking water and 4.2 billion people in the world (54%) were without access to sanitation services in 2020.To overcome the problem, initiatives are being taken at international level to improve access to sanitation through various programmes. In India, various initiatives have been taken to improve access to safe drinking water and sanitation in the country, such as Accelerated Rural Water Supply Programme, National Drinking Water Mission, Rajiv Gandhi National Drinking Water Mission, Jal Jeevan Mission to improve supply of drinking water. The various sanitation programmes like Central Rural Sanitation Programme, Total Sanitation Campaign, Nirmal Bharat Abhiyan, Swachh Bharat Abhiyan were introduced to improve sanitation and hygiene among rural households. There has been significant improvement in access to water and sanitation in the country. However, still good proportion of people are without access to safe water and sanitation in rural area. Further, there are issues of unsustainability of water sources and inadequate hygiene practice.

In North East India, only 46 per cent habitations were covered with drinking water in July 2021. Arunachal Pradesh is a remote State of India with low sanitation coverage and water supply. Sanitation coverage in the State has improved tremendously under Swachh Bharat Abhiyan. But there are issues related to quality of facility, sanitary habits and hygiene practices. The access to safe water is unsatisfactory and there are issues related to solid and liquid waste management in various urban centres. Access to sanitation, among other factors, depends on attitude, knowledge and awareness among the people about various aspects of sanitation and its importance. Having access to sanitation facilities does not fully guarantee proper sanitation, it also depends on sanitary habits and hygiene practices among the people. This book presents the status of access to sanitation, hygiene practices, level of awareness about sanitation and Swachh Bharat Abhiyan among the Scheduled Tribe population and issues related to solid waste management system in Papum Pare district of Arunachal Pradesh.

Bikash Bage
Ashi Lama

Acknowledgements

At the very outset, we would like to acknowledge the financial assistance provided by the Ministry of Social Justice & Empowerment and Ministry of Tribal Affairs, Government of India, New Delhi to carry out a research study on 'Sanitation among the Scheduled Tribes of Papum Pare District of Arunachal Pradesh—A Case Study of Swachh Bharat Abhiyan'. We thank the Department of Social Justice & Empowerment and Tribal Affairs (SJETA), Government of Arunachal Pradesh, Itanagar for extending support to conduct this study. The study would not have been completed without the support of the Department. We would like to express our thanks and gratitude to Professor Saket Kushwaha, Vice-Chancellor of Rajiv Gandhi University for his encouragement in taking up this work. We are also thankful to Dr. N.T. Rikam, Registrar of Rajiv Gandhi University for the administrative support. We also acknowledge the help and support of Shri Yumlam Kaha, Director, SJETA, Government of Arunachal Pradesh and his entire team of officers and officials for extending full cooperation during the course of the study.

We would also acknowledge the help and support of Officers and staffs of UGC Project Cell, Rajiv Gandhi University for their valuable administrative support.

We thank all the field investigators, namely Ms. Nani Onya, Ms. Jumyir Riba and Ms. Tasso Manu and MTS Ms. Nabam Thoru for their sincere and dedicated work in collecting data and necessary information from the households. We would like to acknowledge the support of Mr. Konlom Mossang, Mr. Dambaru Bhattarai and Shri Vijay Kumar Prasad for data entry and data processing work. We express heart-felt thanks to all the respondents, household heads and local public leaders for their valuable support and cooperation during the field survey and focus group discussion.

Finally, we are thankful to our family members for bearing with us and extending moral support and cooperation during the course of the study. Once again, we would express our sincere thanks and gratitude to all those who helped and supported us during the course of the study.

Bikash Bage
Ashi Lama

Contents

List of Tables and Figures

TABLES

FIGURES

1
Introduction

Introduction

Sanitation is vital for living a healthy and productive life. It enhances the quality of life as well as overall socio-economic development of a nation (Srivastav, 2022). It is the most basic human need for good health and well-being (The United Nations). It refers to public health condition related to supply of safe drinking water, access to sanitary latrines, proper drainage and sewage system as well as proper and scientific disposal of solid waste. Access to safe water and sanitation is prerequisite for good health and also for alleviating poverty (Nath, 2017). In fact, the quality of life is largely dependent on access to proper sanitation (Sengupta, 2017). Proper sanitation also improves human productivity, reduces healthcare cost and promotes social and economic prosperity. Water and sanitation are related intrinsically. Adequate supply of safe water and sanitation is very important for women and girl children as the lack of these facilities can expose women to insecurity and health hazards. It can also adversely affect food security, level of nutrition and livelihood choices of women (Cronin, *et al.* 2015). United Nations has declared ensuring access to safe water and sanitation as one of the 17 goals of sustainable development. Though the sanitation is very important for the well-being of people and development of a nation, a large number of people in the world, particularly in developing countries, do not have access to proper sanitation facilities.

The Joint Monitoring Programme (JMP) of the United Nations Children's Fund (UNICEF) and the World Health Organisation (WHO) using the concept of 'safely managed services' (availability, accessibility and quality) estimated that 2 billion people in the world (26%) were without access to safe drinking water in 2020. In rural areas, 60 per cent of the people in the world lacked access to this level of services (Heller, 2022). Regarding sanitation service, the JMP reported that 4.2 billion people in the world (54%) were without access to sanitation services, of which 494 million were still practicing open defecation. In 2019, 31 per cent of schools worldwide lacked access to quality water and 37 per cent of the schools did not have sanitation facility (UNICEF & WHO, 2020). Further, one-third of the health care facilities, in 2020, did not have hand washing facilities. The Joint Monitoring Programme report, "Progress on drinking water, sanitation and hygiene, 2000-2017: Special focus on inequalities", finds that while significant progress has been made towards achieving universal access to Water Sanitation and Hygiene (WASH), still there lies huge gaps in the quality of services provided. The report estimate shows that 1 in every 10 people still lack basic services, including 144 million individuals who drink untreated surface water. While open defecation has been halved since 2000, from 21 per cent to 9 per cent but still millions of people continue this practice in 'high burden' countries. Moreover, in 39 countries, the majority of which are in sub-Saharan Africa, the number of people openly defecating has increased. Every year, 2,97,000 under-age-five children die from diarrhoea linked to inadequate WASH. Poor sanitation and contaminated water also help transmit diseases, such as cholera, dysentery, hepatitis A, and typhoid.

The lack of safe water and sanitation are critical threat to public health. Hence, the global community has been making efforts to improve access of people to water and sanitation. The World Health Assembly in 1978 adopted four key strategies for good health of all people which are: health education for prevention of diseases, promoting healthy lifestyle and reducing risk factors, supply of adequate safe water and sanitation, nutritional promotion including food supply. In order to encourage global community to improve access of people to water and sanitation, the Action Plan of United Nations Conference on water held at Mar del Plata in 1997 states that all people, whatever their stage of development and social and economic conditions be, have the right to access to drinking water in sufficient quantities and of a quality equal to their basic needs. Further, the International Decade on Water Supply and Sanitation was launched in 1980 by the United Nations to provide universal access to quality water services. The United Nations General Assembly on 28th July, 2010 approved the resolution and declared 'right to water and sanitation' as a human right. It called upon the States and international organisations to provide financial resources, capacity-building and technology transfer to ensure supply of safe, clean, accessible and affordable drinking water and sanitation for all (Heller, 2022).

India is a vast country with diverse population. As per 2011 Census, it has a total population of 1.21 billion, out of which 68.8 per cent live in rural areas. The country's literacy rate was 74.04 per cent in 2011 Census. The availability of safe drinking water and sanitation in India had been miserable till the end of 20th century (Sengupta, 2017). However, the water supply and sanitation in the country has greatly improved over the years under the planned development process and implementation of various programmes, such as National Drinking Water Mission (NDWM), Central Rural Sanitation Programme (CRSP), Total Sanitation Campaign (TSC), Nirmal Bharat Abhiyan (NBA) and ongoing Swachh Bharat Abhiyan (SBA). Despite efforts and significant progress, still millions of people in the country, particularly in rural areas, lack access to clean water, toilets, and sewage infrastructure. Various government programmes at national, state, and community level have brought rapid improvements in sanitation and the drinking water supply. These various programmes are ongoing. In 1980, rural sanitation coverage was estimated at 1 per cent and reached 95 per cent in 2018. Also, the share of Indians with access to improved sources of water has increased significantly from 72 per cent in 1990 to 88 per cent in 2008. Life expectancy at birth has doubled since Independence, from 33 years in 1947 to 68 years in 2011. The issues of health and well-being are closely related to that of an adequate water supply and functional sanitation systems. India is focused on ensuring access to water and sanitation services to all. Since the launch of Government of India's flagship scheme, the Swachh Bharat Abhiyan (Clean India Mission), more than 10 crore toilets have been constructed in rural areas. Despite commendable progress, India still faces significant challenges in the provision of quality water, sanitation, solid waste management and drainage. Further, there is high inequality in access to sanitation with more than 80 per cent of urban households accessing sanitation facilities compared to only 30.7 per cent in rural India (Census, 2011). In 2011, 53 per cent of the households in the country did not have access to individual household latrine. Access to sanitation in India had been poor not only due to poverty and low income, but also due to mindset, cultural barriers and superstitious beliefs. This issue was rightly highlighted in the Hindi movie '*Toilet Ek Prem Katha*'. Swachh Bharat Abhiyan aims to change the mindset of the people towards sanitation and make it a habit among the people.

Arunachal Pradesh, the eastern most State of India, is one of the late starter of the development process. During the colonial period, the area was kept as a buffer zone and so only a little developmental activities were taken up. Even after the Independence, the area did not receive much attention. It was called North Eastern Frontier Agency and was administered centrally. The actual process of development started with the grant of Union Territory status on 20th January, 1972 and full-fledge Statehood on 20th February, 1987. As per 2011 Census, the State has a total population of only 13.83 lakh, out of which 77 per cent live in rural area. The literacy rate of the State was 65.38 per cent in 2011 Census.The access to water and sanitation has improved over the years due to efforts made by the government under various plans and schemes. However, still many people in the State lacks access to safe water and proper sanitation. The Census 2011 showed that 38 per cent of the households in the State did not have sanitary latrine. In rural area of the State, 47.3 per cent of the households were without latrine facility. In case of access to safe water, 34 per cent of the households lacked access to tap water and in rural area 41 per cent of households were without access to tap water. Access to sanitation depends on awareness, knowledge and attitude of the people apart from income and financial resources. Among the 25 districts of the State, Papum Pare district is the most populated district. It has a total population of 1.76 lakh (12.7% of total population of the State). This district is inhabited by almost tribes of the State as it houses State Capital. In this background, the study was carried out to examine the status of access to sanitation among the Scheduled Tribes of Papum Pare district of Arunachal Pradesh.

Statement of the Problem

The progress of a nation mainly depends on its public health. The public health is its greatest asset. Therefore, no compromise should be made on it. Poor sanitation is the greatest threat to public health. India is recognised as the second populous country after China. But India is lagging far behind in terms of sanitation. India is followed by China and Africa in respect of sanitation where majority of its people are without adequate sanitation. Sanitation is the most neglected sector in India. Even though responsibility of providing adequate sanitation and water supply rests with the respective State Governments, the State Governments had neglected this sector from the recent past. Union Government emphasizes on investments in other developmental sectors of the country but fails to recognise the need of adequate sanitation for its citizens. Such negligence affects the well-being of the people of the country. People in India have greater interest in procuring luxurious assets; like cell phones but fails to realise the necessity of toilets and sanitation. Inadequate sanitation threatens the health and hygiene of people.

Majority of Indians are still leading a life in unhygienic condition. In India, nearly 626 million people practice open-defecation which is equivalent to 60 per cent of open defecation in the world.

Sanitation is a global issue. It is the basic necessity of life. It is an unrecognised basic need of people. The proverb "Health is Wealth" is necessarily true. Adequate sanitation not only promotes a healthy living but also leads to economic development of a nation. It decreases the cost of medical expenditure thereby increasing savings of people. This increase in savings is an indicator of economic growth of a nation. The total economic impact due to inadequate sanitation in India is estimated at an amount of Rs. 2.44 trillion (US$53.8 billion) a year, which is equivalent to 6.4 per cent of India's GDP in 2006 (Water and Sanitation Programme, 2007). Massive water and soil pollution leading to increase in school drop-outs, loss of income from tourism and other sectors are other adverse effects of inadequate sanitation.

Access to water and sanitation is considered as a human right. It also comes under the provisions of Article 21—Right to Life, which is a fundamental right guaranteed by the Constitution of India. Recognising the importance of right to water and adequate sanitation, the UN General Assembly in July 2010 has given due importance in placing it amongst other rights given to its citizens. Clean drinking water and improved sanitation are key drivers in poverty reduction and leads to the realisation of all human rights (UNDP, 2011). The success of such initiatives is based on successful planning and proper implementation of water and sanitation related projects. November 19th is celebrated every year across the world as World Toilet Day to create awareness and remove unawareness about the basic sanitation currently affecting 2.5 billion people. The year 2008 was declared as International Year of Sanitation.

The growth in urban areas in India is faster than rural areas. The population growth in urban areas in India has increased from 27.8 per cent in 2001 to 31.8 per cent in 2011. The number of towns is also increasing rapidly. But these urban areas are lacking in adequate sanitation facilities and thus the majority of urban population in India could not lead decent standard of living. The rapid increase of population in urban areas has resulted in poor health outcomes. As such the Government of India in October 2008 announced the "National Urban Sanitation Policy" (NUSP) to improve sanitary condition in urban areas. Tribal group is a socially defined category of people who identify with each other based on shared social experience or ancestry. India has the second largest tribal population in the world, next only to Africa. Over 84 millions of India's people belong to Scheduled Tribes. Tribal population makes up 8 per cent of India's population. They are also called hill tribes or primitive tribes. According to 2011 Census, around 27 per cent tribal population do not have access to safe source of drinking water and 75 per cent of tribal households in the country do not have toilets and are defecating in the open. Acting in accordance with 'Swachh Bharat Abhiyan' launched by Government of India in 2014, Arunachal Pradesh Government had targeted to achieve cent-per-cent open defecation free (ODF) status for all districts by December 2017. Arunachal Government has initiated cleanliness campaign "*Hamara Elan Swachhta Aur Nari Uthan*" to encourage citizens to keep their surroundings clean as well as to empower women. A cleanliness protocol has been also developed by Arunachal Pradesh Government. The Census 2011 showed that 47.3 per cent of rural households in Arunachal Pradesh were without latrine facility. In case of access to safe water, 41 per cent of rural households were without access to tap water. Hence, the study was undertaken to examine the status of access to sanitation, sanitary habits and hygiene practices and awareness about Swachh Bharat Abhiyan among the Scheduled Tribes of Arunachal Pradesh taking Papum Pare district as a representative district.

Review of Literature

There exist a number of literatures on Sanitation and its allied issues. In this section keeping in view of the objectives of the present study, a brief review of relevant literature has been presented:

In the article, "A health profile of tribal India", S.K. Basu (1994), stated that modern medicine has not been accepted in most tribal areas, where magico-religious health care systems prevail. Health conditions in tribal areas have been described as deficient in sanitary conditions, personal hygiene, and health education. Health workers need information on the folklore related to health of tribal groups, so that they could provide appropriate health and sanitary practices and to document indigenous herbs for medical use. Malnutrition is pervasive among tribal population. The workload of tribal women is heavy, long and increasing. Maternal mortality is due to unhygienic conditions and inappropriate tribal practices. Root, G.P.M. (2001) in the article "Community Environments and Childhood Diarrhea in Rural Zimbabwe deals with a comparative study on two communities – one having access to latrine and the other without a latrine. The author tried to link access to sanitation facilities with diarrhoeal morbidity and diarrhoeal mortality.

S.K. Jana and Pradip K. Bhowmick, in the article "Rural Sanitation Programme and Tribal Health: A Study in the District of Midnapore, West Bengal" (2003) stated that the problem is not unique to West Bengal. Under such circumstances, the task of implementing a sanitation programme is not an easy one because the challenge lies in changing mindsets of the people concerned more than changing habits. The study conducted among the tribal people of Midnapore district in West Bengal through primary and secondary data, finds that the majority of the tribal people are unaware about the concept of proper sanitation and safe drinking water. The people belonging to general higher caste groups are better aware about the personal hygiene and environmental sanitation and safe drinking water than the tribal people. That is why the higher caste people, who happened to be better off, both educationally and economically, are better aware about the significance of personal hygiene and environmental sanitation and safe drinking water.

In the article "Community Sanitation Campaign: A Study in Haryana" (2008), Gupta, V. and Mahi Pal made an attempt to evaluate the processes and outcome of the implementation of

Community-Led Total Sanitation (CLTS) approach. They argued that ill health only due to poor sanitation results in high expenditure and loss of savings. Poor sanitation results not due to poverty, but also due to unawareness on sanitation among the people. The study also compared the situation prior to introduction of CLTS and the situation after its implementation. They viewed that CLTS has motivated villagers for sanitary practices. The introduction of CLTS approach has lowered the health expenditure and improved the health conditions of the people.

Anbarci, Escaleras and Charles A. Register in the article, "The III Effects of Public Sector Corruption in the Water and Sanitation Sector" (2009), have considered access to drinking water and sanitation as a necessity not luxury. The diseases occurring due to lack of sanitation and drinking water have impact in the lives of the sufferer in their later lives also. The problem of inadequate sanitation and water is not only limited to developing countries, but is spread across the world. Public sector intervention is needed to meet the expenditures required to improve the sanitary condition. They believed that the involvement of Government through private parties opens the gateway to corruption. Corruption retards the growth of the system giving accessibility of improved drinking water and sanitation to only a small portion of the total population. The authors viewed corruption as an inseparable part from water and sanitation. They found water and sanitation to be the most corrupted sectors worldwide. Corruption starts at the construction level which involves public-private actors. The entire study was based on the secondary data.

B. George in the article, "Sanitation Programme—a Glass Half-Full" (2009), has focused on the positive aspects of Nirmal Gram Yojana thereby criticizing the view given by Malhotra and Patnaik in their article about the maladies associated with Nirmal Gram Yojana. The role of community participation in Total Sanitation Campaign (TSC) has been emphasized. Poor sanitation is linked with unhealthy living. Poor sanitation also affects investments in education thereby affecting gender disparity in schools. Due emphasis is also given in adopting strategy to ensure sustainability of sanitation through community participation and awareness. The study argued for the greater participation of women for improving sanitation.

In the article "Combining Sanitation and Women Participation in Water Supply: an Example from Rajasthan" (2010), K.O. Reilly viewed that public interest in sanitation is lower than interest in water supply. People are not conscious of their health. They are more interested in accessing water than constructing latrines at home. Women are considered as the main beneficiaries of latrine for their safety, privacy and status. Therefore, their participation is often considered as a part of an efficiency approach to sanitation. Women are considered as the introducers of hygiene practices in their homes. Male members of the family should feel it as their responsibility to provide safety and privacy for the women in their family by constructing latrines.

Shakuntala C. Shettar in the article, "Sociology of Sanitation: Incorporating Gender Issues in Sanitation" (2013), emphasized on certain gender issues related to sanitation. Gender biasness is reflected in maintaining proper sanitation. The role of both men and women varies in sanitary activities which is more prominent in rural areas. Women are considered as sole agent of creating sanitary environment in the environment. They are worst sufferer in times of water crisis and improper sanitation. Women are responsible for maintenance of hygiene in their family. The responsibility of the health condition of the family rests with the women. The prevailing gender ideology has affected the status of women. The gender ideology also imposes restrictions on women in terms of using toilets. The study provided certain recommendations for improving sanitation programmes. The line ministries such as health ministry, water resource ministry etc. need to address gender issues in sanitation policies and legal framework. The community leaders should be targeted for gender sensitisation. Due consideration and fund allocation in school sanitation and hygiene programmes is required as school children are influential agents of hygiene consciousness in the families. The structural design of the latrines should cater to the special needs of the girl child in the schools. Sanitation designs in schools should also cater to the physically challenged students in the schools.

Robert Chambers and Gregor Von Medeazza in the article, "Sanitation and Stunting in India under nutrition's Blind Spot" 2013, draws the relationship between under nutrition and open defecation. The effects of open defecation is found more on children below the age of five. They stressed on the child under-nutrition instead of focusing on the impact of open defecation and insanitary practices associated with under-nutrition. The study stressed on other relevant factors like feeding programmes, the amount and quality of food intake etc. They viewed that the problem of high population density further worsens the problem of open defecation. The rate of open defecation is still high.

Mohammad Akram in the article "Sanitation, Health and Development Deficit in India" (2013), argued that the individual, the community as well as the State are jointly responsible for Sanitation in both personal and public life. The goals of public health could be achieved through Sanitation. Development deficits are mainly responsible for insanitation in India. Sanitation deficits are promoted by bureaucratic targetism, medicalism, povertism and dehealthism. Urban slums and rural inhabitants are mostly affected by insanitation. The concept of health varies among social groups. The health policies and programmes in India had certain deficiencies. Open defecation is the indicator of development deficit. It is misconceptualised as a cultural deficit. People are unnecessarily blamed to conceal the deficiencies in policy making and policy implementation. Sanitation in India is yet to become an integral part of development deficit. The problem of open defecation not only accounts for rural or urban poor class but also the elite and the governing class.

In the article, "Gender Perspectives on Sanitation for Sustainable Development", Begona Lasagabaster (2013), stated that women are affected by inadequate sanitation due to many physiological, social and cultural factors. Women are the main sufferers of reproductive tract infections due to poor sanitation. Improper sanitation limits their freedom of mobility. They are kept away from participating in decision making and other community affairs. The development planners, the water and sanitation community and the violence against women community must recognise these challenges of women. It also requires the attention of the local leaders and community members involved in deciding the facilities to be provided for the people. Some other policy actions include consultation with gender-based violence specialists, designing, constructing and managing infrastructure for reducing vulnerabilities, involving the women section into the planning process. Addressing the immediate needs of women in water and sanitation sectors should be accompanied by addressing violence against women.

In the article, "Causes of Insanitation in India and Ways of Improvement" (2014), Mohammad Akram viewed that the perception of sanitation varies from countries to countries. The vulnerable and the marginalised section of society are mostly affected by inadequate sanitation. The modern developed countries consider Sanitation as 'public good'. But sanitation as a 'public good' is not represented in policy making. The concept of sanitation has extended from cleanliness and hygiene to collective aspirations relating to liberty, equity, freedom and social justice. Sanitation as an issue has not attracted the reformers, policy makers as well as administrators. Insanitation is deeply rooted in the socialization process. The sociologists claim toilet training as the important goal of early socialisation. Lack of social sanctions related to insanitary behaviour in societies like open defecation has added momentum to the problem of insanitation. Further, the prevailing notion of pollution and purity in societies associated has led to the growth of insanitary practices. The avoidance of upper castes in sanitary activities and involvement of only the so-called scavengers in the sanitary activities is responsible for insanitary practices.

Trisha Agarwala in the article, "Urban Sanitation in India: A Growth Story Gone Awry" (2015), has expressed her view that water supply and sanitation is the most neglected sector in India. The investments in these sectors seem to be discouraging. The study argued that the city authorities have to play pivotal roles in sewage treatment, solid and liquid waste management. It also emphasized the positive effects of sustainable sanitation among the women and girls. The study stressed on the necessity of a more serious campaign towards sanitation to stop manual scavenging. New technology should be employed for toilets. The study viewed that all sorts of policies and schemes would be fruitless unless backed by required budget and implementation.

Poornima Chikarmane in the article, "Swachh Bharat Abhiyan: Marginalising the Mainstream" (2015), stated that any sort of cleanliness campaign must give due dignity and justice to the workers involved in the waste handling systems. The study asserted that the works of the informal recycling workers are skilled works and due recognition should be given to their works. They are the core workers in the waste handling system.

Bhasha Singh in the article "Sanitation, Development and Social Change: Cleaning of Holy Minds" (2015), expressed his views on changing the mindsets of people on the sanitation workers. He has discussed how the scavengers are looked down by people. He has also expressed about the pitiable condition of the lives of the scavengers. He felt that due status and dignity should be given to them and the services that they provide. He felt that social change cannot be achieved unless we learn to give due honour to their work. The chain between the caste and scavenging works should be broken if real development is needed.

The article "Swachh Bharat: Some Issues and Concerns" (2015) by P.V. Subbarao and S. Somasekhar, deals with reviewing the earlier sanitation initiatives and the lessons that could be learnt from those failures. They have given due significance to the effects of open defecation, manual scavenging in the article. Gandhi's views on sanitation and health have been described in it. They have cited the challenges of poor sanitation in the society as well as suggested remedies to resolve the problems. They stressed on changing the mindset of people and change in behaviour in respect of practicing open defecation. They held the view that the agenda of sanitation in modern times must cover total sanitation all over the country instead of focusing on villages, towns or cities. They insisted on widening the coverage of sanitation from sanitary toilets to sustainable sanitation and clean environment.

Cronin *et al.* (2015) argued that gender equality is central and critical to economic and human development. But in developing countries gender disparity in access to water, sanitation and hygiene is high. It was stated that women have primary responsibility in managing household water supply, but they are hardly involved in water management in policy discourse in water management.

In the article entitled "Status of Sanitation and Hygiene Practices in the Context of Swachh Bharat Abhiyan in two Districts of India" (2016), Puspanjali Swain and Sristhi Pathela have mainly emphasized on urban slum and rural inhabitants of India as they are neglected in terms of basic sanitation facilities. The obstacle which hinders the success of Swachh Bharat Abhiyan has been sought out. They suggested that community participation is needed for the success of Swachh Bharat Abhiyan. Behavioural change is expected to solve sanitation related problems. Pressure of population on the nation has made the problem of sanitation more complex.

Tomar, N.S. in the article "Swachh Bharat Mission: From Behaviour Change to Social Change" (2016), has expressed the view that Swachh Bharat involves mainly behavioural change. He said that this Mission has aimed in removing the hindrances of availing clean environment. He spoke about the positive sides of the mission. The role of media in this regard has also been appreciated. Various components of the Mission have been described in brief.

In the article "Understanding Open Defecation in Rural India—Untouchability, Pollution and Latrine Pits", D. Coffey, A. Gupta, P. Hathi, D. Spears, N. Srivastav, S. Vyas (2017), have compared the rural sanitation scenario of India with other countries. They explored the social causes of purity and pollution to Open Defecation. They made an assumption that the underlying cause of Open Defecation may be beliefs, values, norms, purity, pollution, caste and untouchability. They are of the opinion that the misconception of manually cleaning the pit latrines by the Dalits has increased the rate of open defecation in villages. They viewed that poverty and water inadequacy cannot be blamed for open defecation in India. They expressed their opinion that to make India open defecation free, the future rural sanitation policy must emphasize on removing the biasness of pit latrines from rural Indians.

Mohan, R.V.R. in the article, "Swachh Bharat Mission (Gramin)—Bottlenecks and Remedies" (2017), has identified the barriers to success of Swachh Bharat Mission by 2019. He has also provided certain remedies to meet the obstacles. He viewed that the Swachh Bharat Kosh instead of spending funds on defunct toilets and water supply on the doorsteps of toilets has misused funds. The non-availability of water at the doorsteps of the toilets has made the problem more complex. He is of the opinion that the availability of water supply at the door steps of the toilets with Individual Household Latrine promotion will make the women members of the community more pleased. The initial stage expenditure in construction of latrines under Swachh Bharat Mission could be provided as loans by the Women Self Help Group's, Non-Governmental Organizations etc. He insisted on making functional the defunct toilets through some new initiative.

In the article, "Drinking Water, Sanitation and Waterborne Diseases" (2017), Rashmi Tiwari and Sanatan Nayak have drawn a relation between safe drinking water and sanitation with health. They have analyzed through multi-stage sampling technique the different sources of drinking water in the four selected zones. The study reflects that accessibility to safe drinking water and improved sanitation is correlated with education, income levels and caste. High level of education, high level of income and upper caste people have access to safe drinking water and improved sanitation in contrast to low income earners, less educated people and people belonging to low caste. As the low income earners, less educated and low caste people have to rely on piped taps, they are more prone to diseases than the others. They have also suggested that the best way to prevent waterborne diseases is through purification of drinking water. In this article the authors have overemphasized on drinking water and have given less emphasis on sanitation facilities and water-borne diseases.

Gopinath Pradhan in the article "Hygienic Practices among Tribal Communities: Case Study of Odisha" (2015), attempted to understand the awareness, knowledge and practices of tribal with respect to overall sanitation, hygiene and drinking water. The awareness among the community regarding sanitation and handling of drinking water is not adequate. The women and girl children are primarily responsible for collection, storage of water for household use and maintaining hygiene and sanitation. NGOs of the locality can also play a vital role in improving the safe drinking water supply and sanitation facilities of the rural households through creating awareness.

Kumar *et al.* (2022) pointed out insecurity of drinking water in rural India. The study found that coverage of drinking water supply in rural India has increased significantly due to efforts made by the government under various plans and programmes. By the end of July 2021, 79 per cent habitations in rural India were covered by water supply. But still 59 per cent of rural households in India as in July 2021 were without functional pipe. It was due to unsustainability of water sources, poor cost recovery, high operation and maintenance costs. There are other issues like inequality in access to water and unsafe source.

The review of literature shows that there is a dearth of study which have explored the status of sanitation among the tribal population of Arunachal Pradesh in a comprehensive manner. Hence, the present study was carried out to explore the status of sanitation, awareness and attitude towards sanitation, awareness about Swachh Bharat Abhiyan (SBA) as well as hygiene practices among the Scheduled Tribes of Arunachal Pradesh by taking Papum Pare district as a representative sample district.

The Objectives of the Study

The objectives of the research are as follows:

1. To assess the status of sanitation and basic hygiene practices among the Scheduled Tribes of Papum Pare district of Arunachal Pradesh.
2. To identify the problems related with poor sanitation among the Scheduled Tribes in Papum Pare district of Arunachal Pradesh.
3. To examine the level of awareness about Sanitation and Swachh Bharat Abhiyan among the Scheduled Tribes in Papum Pare district of Arunachal Pradesh.
4. To evaluate the system of solid waste management in Papum Pare district of Arunachal Pradesh.

Data Source and Research Methodology

The study aims to assess the sanitation and hygiene among the Scheduled Tribe population. The study was conducted in Papum Pare district of Arunachal Pradesh. The justification for the selection of this specific area is that it is the most populous district of the State and is inhabited by almost all the tribes of the State. As per 2011 Census, this district has a population of 1,76,573 accounting for 12.76 per cent of the total population of the State. The population density of this district was 51 persons per sq. km. The rate of migration to the towns of this district has also increased rapidly. Administratively, the district is divided into three sub-division, namely Itanagar Capital Complex, Yupia and Sagalee. It is further divided into 15 administrative circles.

The study will rely on primary and secondary data. Since secondary data on sanitation are inadequate, it will generate primary data through extensive field survey. The secondary data were collected from the 2011 Census report, Ministry of Drinking Water and Sanitation, Statistical Abstracts of Arunachal Pradesh, journals and books.

The study was based on Multi-stage sampling technique. In the first stage, Papum Pare district of the State was selected on the basis of the size of population. In the second stage, 7 circles of the district were selected on the basis of distance from the district headquarters. In the third stage, from each circle four to five villages were selected for the study. In total, 29 villages were covered for the survey and data collection. A mix of remote and well accessible villages were selected to make the sample representative. Finally, from each village 20 to 30 households were selected randomly for data collection. The total sample size was 695 households.

Table 1.1: The Block/Circle-wise distribution of sample households

Block/Circle	No. of household surveyed	Percentage
Doimukh	125	18
Balijan	95	14
Sagalee	120	17
Kimin	59	9
Itanagar	100	14
Naharlagun	124	18
Banderdewa	72	10
Total	**695**	**100**

Source: Statistical Abstract of Arunachal Pradesh-2015.

Table 1.2: Village-wise distribution of sample households

Block/Circle	Village	Sample size
Doimukh	Rono Village	25
	Amba	25
	Emchi	25
	Huto	25
	Mani	25

(*Contd.*)

Block/Circle	Village	Sample size
Balijan	Hollogi	25
	Kokila	16
	Upper Balijan	27
	Upper Tubung	27
Sagaleee	Chumbang	20
	Kamlee	30
	Lower Gai	30
	Nyariso	20
	Yallang	20
Kimin	Durpa	24
	Model Village	25
Itanagar	Chimpu	25
	Ganga	25
	Jollang	25
	Kongko	25
Naharlagun	Borum	25
	Sood	24
	Nyorch	25
	Lekhi	25
	Papu Nallah	25
Banderdewa	Karsingsa	25
	Duda	15
	Tani	11
	Tarajuli	21
Total	**29 villages**	**695**

Source: Statistical Abstract of Arunachal Pradesh-2015.

The primary data were collected by conducting field survey in the selected villages. The survey was conducted by the field investigators with the help of well-structured schedule. The direct personal interview method was used to collect the required information from the households. The schedule was designed to collect information on socio-economic profile of households, knowledge and attitude towards sanitation, access to sanitation, sanitary habits and hygienic practices, sanitary status in locality, knowledge and participation in Swachh Bharat Abhiyan and problems in accessing sanitation. Both closed and open ended questions were prepared to collect the information. Since the common language of the area is Hindi, therefore, the questions need to be translated into Hindi while interviewing. From each household only one respondent was selected for interview and collection of necessary information and data. The unit of study was household. The respondents were preferably the head member of the household or his/her representative. In addition to interview, data were also collected through personal observation of the researchers. Further, focus group discuss with key informants were conducted to collected the relevant information. The data were analysed using various Statistical tools such as frequency distribution, descriptive statistical methods, line and bar graphs and cross tabulations.

Profile of the Study Area—Papum Pare District

Papum Pare district is spread over an area of 3,462 sq. km. The district is administratively divided into five community development blocks, namely Doimukh, Sagalee, Kimin, Balijan and Mengio. It is further divided into 15 administrative circles, namely Balijan, Itanagar, Naharlagun, Doimukh, Toru, Sagalee, Leporiang, Mengio, Kimin, Banderdewa, Tarasso, Kakoi, Gumto, Parang and Sangdupota. As per 2011 Census, the district had a population of 1,76,563 persons accounting for 12.76 per cent of the total population of the State. The total Scheduled Tribe population in the district was 1,17,216 persons (66.38% of total population). It had a population density of 51 persons per sq.km in 2011 against the State average of 17 persons per sq. km.

Table 1.3: Profile of Papum Pare District of Arunachal Pradesh (2011 Census)

1	Area (in sq.km)	3,462
2	Population	1,76,563
3	Population density (persons per sq. km)	51
4	Urban population (%)	55
5	**Percentage of ST population**	**66.38**
6	Literacy rate (%)	79.95
7	Sex ratio	980
8	Number of villages	485
9	Number of households	35,730

Source: Statistical Abstract of Arunachal Pradesh-2015.

The literacy rate of the district was 79.95 per cent which was much higher than the State average of 65.38 per cent. The sex ratio of the district (980 females per 1000 males) was above the State average of 938 females per 1000 males. It is also the most urbanized district of the State with 55 per cent urban population. It is the most cosmopolitan district of the State as it is inhabited by almost all the tribes of the State and various communities from other States of India by virtue of housing the twin capital towns, namely Itanagar and Naharlagun.

Significance and Relevance of the Study

Sanitation is essential ingredient for improving public health condition. Swachh Bharat Abhiyan is a cleanliness mission for the betterment of the life of the people in India. It is seen that every year the death rate increases especially among the infants and children due to unhygienic living condition. They are prone to many diseases relating to worms, skin infections, water borne diseases like malaria, diarrhoea, jaundice, urinary tract infections etc. Swachh Bharat Abhiyan has mainly emphasized on open defecation free (ODF) India. Open defecation has become a burning issue in our country. It can be eliminated only through provision of proper sanitation facilities and changing the behaviour of people towards sanitation. Poor sanitation has also resulted in increase in number of school drop-outs especially among the girls. In certain cases, people practise open defecation not due to insufficient fund to build toilets, but because of an attitude deeply rooted in their social and personal habit. Therefore, people should be made to understand the importance of using toilets and realise the danger of open defecation. Solid and Liquid Waste Management is another important component of the programme. Improper management of wastes also threatens public health and adversely affects flora and fauna. Improper waste disposal can lead to serious diseases. The problem of waste disposal is more in urban areas than in rural areas due to lack of space. Thus, the sanitation goals cannot be achieved, if waste is not safely disposed of.

Sanitation is a serious problem among tribal groups. They suffer from diseases like forest fever, yellow fever, typhoid, malaria etc. The cultural attitude of the tribal people sometimes becomes problem in establishing sanitation facilities. Thus, Sanitation is a key to good health and it helps in betterment of the human resources of a country. The status and access to sanitation among the tribal population of the country has been poor due to various reasons ranging from low income, lack of awareness and socio-cultural barriers. This book throws light of on various aspects of sanitation among the Scheduled Tribes of Papum Pare district, Arunachal Pradesh. The book is expected to be useful for policy makers of the State in designing strategies to overcome the lacunas and improve access to quality sanitation facilities among the tribal people. It is also expected to be useful to academicians, researchers and students.

2

Sanitation in India, Arunachal Pradesh and Papum Pare District

An Overview

Introduction

Sanitation is essential for good health and well-being of the people. The main components of sanitation consist of access to sanitary latrine or toilet, access to safe drinking water, proper drainage, sewage system and facilities for proper disposal of solid waste. Sengupta (2017) argued that quality of human life is largely dependent on the availability of adequate safe drinking water and proper sanitation. Nath (2017) stated that providing safe drinking water and sanitation to the community is a basic precondition for improving health and alleviating poverty. The lack of access to sanitation facilities can contaminate surroundings and expose the people to various kinds of diseases leading to morbidity and mortality. Thus, sanitation plays a pivotal role in socio-economic development of a community and a nation as a whole. Since drinking water is the basic necessity of life, the Government of India has taken various measures to supply drinking water to the people in rural areas since Independence. In the initial years, the focus was put on providing physical infrastructures, like wells, handpumps etc. However, to ensure sustainability of the system steps were taken to encourage community participation in planning and implementation, operation and maintenance of rural drinking water schemes (Chand, 2017).

The Government of India started Accelerated Rural Water Supply Programme in 1972-73, in order to improve coverage of villages with drinking water supply. Under this scheme, the Central Government provided financial and technical assistance to implement water supply schemes. The Government of India launched National Drinking Water Mission (NDWM) in 1986 to further accelerate the coverage of villages with drinking water and resolve the problem of drinking water. The mission scheme was renamed as Rajiv Gandhi National Drinking Water Mission (RGNDWM) in 1991. Its aim was to cover all rural habitations with population of 100 and above, specially uncovered habitations and water quality-affected areas with water supply and set up water quality monitoring and surveillance system. There was a significant improvement in provision of drinking water in the country. Further, Jal Jeevan Mission was launched in 2019 to provide functional households tap water supply in every rural households by 2024. In April 2005, 96.13 per cent of rural habitations were fully covered (FC) with drinking water, 3.55 per cent were partially covered (PC) and 0.33 per cent were uncovered (NC). But 2011 Census indicated that only 75.71 per cent of the households in India had access to drinking water within the reach of 500 m (Chand, 2017). The estimates of Ministry of Drinking Water and Sanitation indicated that in July 2021, 79 per cent of rural habitations were covered with drinking water (Kumar *et al.*, 2022). But there are issues like slippage of FC habitations into NC or PC resulting from lowering of ground water table, failure of sources and increase in habitations (Chand, 2017). In North East India, only *46 per cent habitations* were covered with drinking water in July 2021. This shows that there is a further need for boosting investment in water supply to enhance coverage and ensure water sustainability in rural areas of the country.

Another important component of sanitation is access to proper sanitary latrine. It is essential for every households to have a proper latrine to avoid human contact to faeces, prevent contamination of water sources and check spread of pathogens. However, in developing countries like India a large number of households lack of sanitary latrine and defecate in open posing threat to health. As per 1991 Census, only 9.48 per cent of rural households in the country had access to individual household latrine. This is because until 1986, rural sanitation was never a priority in India (Sengupta, 2017). Thereafter, the Government of India took certain steps to increase the coverage of sanitation in the country, particularly in rural areas. The Government of India initiated Central Rural Sanitation Programme (CRSP) in 1986 under which latrines were constructed at village premises. But programme suffered due to inadequate allocation of fund. In 1999 the CRSP was converted into campaign mode and it was renamed as Total Sanitation Campaign (TSC). This programme put emphasis on increasing awareness among the people in rural areas about sanitation and generate demand for sanitary facilities. Under it financial incentives for building latrines were given only to below poverty line households. It also provided a toilets in all educational institutions and Anganwadi centres and sanitary complexes at public places. To improve cleanliness of rural environment, the programme also included solid and liquid waste management.

In order to encourage panchayat to promote sanitation, the Government of India in 2004 introduced Nirmal Gram Puruskar (NGP). In July 2012, the government reported that coverage of rural households to individual household latrines has increased to 70 per cent. But Census 2011, reported that only 30.7 per cent households had access to individual household latrines in rural India. To revamp the rural sanitation in the country, therefore, the Government of India launched Nirmal Bharat Abhiyan (NBA) in July 2012. The programme was aimed at addressing the drawbacks ofTSC and to improve the sanitation coverage in rural India. This programme stressed on universal coverage of both below poverty line (BPL) and above poverty line (APL) households, disadvantage households like Scheduled Castes and Scheduled Tribes, small and marginal farmers, landless labourers, physically challenged and women headed households. It provided for financial incentives for building individual household latrines. The newly formed Government of India launched Swachh Bharat Abhiyan (SBA) on 2nd October, 2014 with a view to make the country open defecation free and bring about a behavioural change among the people. The SBA is not fundamentally different from the NBA launched by the previous government (Nath, 2017). However, it puts more focus on creating awareness, information, education and communication apart from financial incentives. It also focuses on the construction of community sanitary facilities and scientific management of solid and liquid waste for better environmental sanitation. The programme is expected to resolve the problem of sanitation in the country.

In this background, this chapter presents an overview of access to sanitation in India and Arunachal Pradesh with focus on Papum Pare district of the State.

Availability of Sanitation Facility in India

Sanitation is recognised as one of the most important issues in India. Historically, the access to sanitation among the people of India was low. During the colonial period, no systematic efforts were made to improve the sanitation facilities until 1860. Since 1860s, municipalities and boards began to be established and measures were taken for improvement of public health. At the time of independence, only 6 per cent of the country's population had access to protected drinking water supply and access to proper latrine were extremely rare (Jha, 2015). However, the sanitation facility in India has improved over the years due efforts made by the government under various plans and programmes, increase in awareness among the people and efforts made by various non-government organisations like Sulabh International Social Service

Organisations which developed and promoted technology for construction of low cost latrines (Srivastav, 2022).

The most reliable data source for sanitation coverage is the Census reports. The improvement in access to sanitary latrine in India as per 2001 and 2011 Census is presented in Table 2.1.

Table 2.1: Availability and Type of Latrine Facility in India

Sl. No.	Type of Latrine	Percentage of households		Change (2011-2001)
		Census 2001	Census 2011	
1	Water closet	18	36.4	18.4
2	Pit latrine	11.5	9.4	-2.1
3	Other latrine	6.9	1.1	-5.8
A	**All types of latrine (1+2+3)**	**36.4**	**46.9**	10.5
B	**No Latrine**	**63.6**	**53.1**	-10.5
	Total (A+B)	**100**	**100**	

Source: Census of India, 2011.

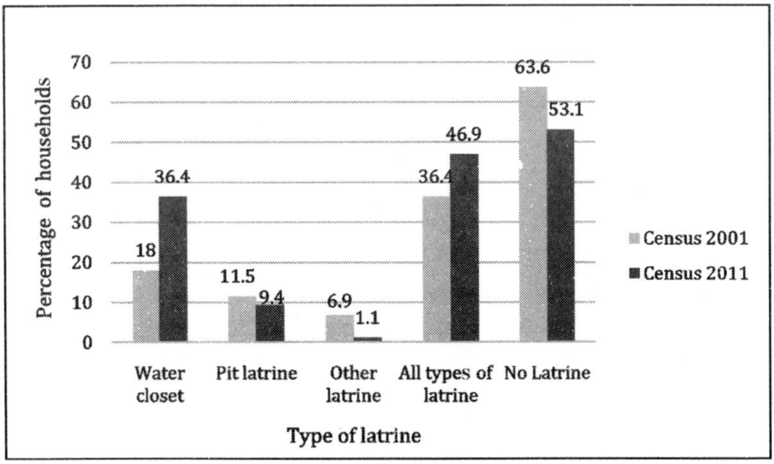

Fig. 2.1: Availability and type of latrine in India.

Table 2.1 shows that access to latrine in the country has improved from 36.4 per cent households in 2001 to 46.90 per cent in 2011. However, despite significant improvement more than half of the households (53.10 %) were without latrine facility. This clearly shows poor state of affairs in the country as far as sanitation is concerned. The analysis of type of latrine showed that in 2001 only 18 per cent of the households had water closet latrine, which improved to 36.4 per cent in 2011 (Figure 2.1). This implies that access to proper latrine in the country is quite low.

The analysis of rural and urban disparity in access to sanitary latrine revealed that in 2001 Census, only 21.9 per cent of the households in rural India had sanitary latrines and 78.10 per cent had no latrine. But in urban India 73.7 per cent of the households had latrine facility and only 26.3 per cent of them were without latrine. In 2011 Census, access to sanitation in rural India had improved to 30.7 per cent households and in urban India, it had improved to 81.4 per cent. The details are given in Table 2.2 and Table 2.3.

Table 2.2: Availability and Type of Latrine Facility in Rural India

Sl. No.	Type of Latrine	Percentage of households		Change (2011-2001)
		Census 2001	Census 2011	
1	Water closet	7.1	19.4	12.3
2	Pit latrine	10.3	10.5	0.2
3	Other latrine	4.5	0.8	-3.7
A	**All type of latrine (1+2+3)**	**21.9**	**30.7**	**8.8**
B	**No Latrine**	**78.1**	**69.3**	**-8.8**
	Total (A+B)	**100**	**100**	

Source: Census of India, 2011.

During the period 2001 and 2011, the percentage improvement in access to sanitary latrine was slightly higher in rural India than in urban India. In rural India, percentage of households having latrine increased by 8.8 percentage points, but in urban India, it increased by 7.7 percentage points. But if quality aspects is considered, it is found that improvement is much better in urban India as compared to rural India. During the same period, in rural India the percentage of households having water closet latrine increased by only 12.3 percentage points, but in urban India it increased by 26.5 percentage points. Thus, rural households of India are having not only low access to latrine, but their access to proper latrine is also poor.

Table 2.3: Availability and Type of Latrine Facility in Urban India

Sl. No.	Type of Latrine	Percentage of households		Change (2011-2001)
		Census 2001	Census 2011	
1	Water closet	46.1	72.6	26.5
2	Pit latrine	14.6	7.1	-7.5
3	Other latrine	13	1.7	-11.3
A	**All type of latrine (1+2+3)**	**73.7**	**81.4**	7.7
B	**No Latrine**	**26.3**	**18.6**	-7.7
	Total (A+B)	**100**	**100**	

Source: Census of India, 2011.

The analysis of status of access to sanitation facilities in selected States and Union Territories of India showed that in 2011 Census, Lakshadweep topped the list with 97.8 per cent of the households having latrine facility followed by Kerala (95.2%) and Mizoram (92%). The details are given in Table 2.4.

Table 2.4 shows that in 2011, Odisha, Jharkhand and Bihar had the lowest percentage of households having access to sanitation. In Odisha and Jharkhand, only 22 per cent of the households had sanitary latrine facility and in Bihar only 23.1 per cent of households had sanitary latrine facility.

Table 2.4: Status of Access to Sanitation Facilities in Selected States of India

States	Percentage of households having access to toilet			No Latrine (%)
	Rural	Urban	Total	
Lakshadweep*	98.1	97.7	97.8	2.2
Kerala	84.3	96.5	95.2	4.8
Mizoram	84.6	98.6	92	8
Delhi*	76.3	89.8	89.6	10.4
Sikkim	84.1	95.2	87.2	12.8
Himachal Pradesh	66.5	89.1	69.1	30.9
Haryana	56.1	89.9	68.6	31.4
Assam	59.6	93.7	65	35
Arunachal Pradesh	52.6	89.6	62	38
West Bengal	46.7	85	58.9	41.1
Gujarat	33	87.7	57.4	42.6
Maharashtra	38	71.3	50.2	49.8
Tamil Nadu	23.2	75.1	48.3	51.7
Uttar Pradesh	21.8	83.1	35.7	64.3
Bihar	17.6	69	23.1	76.9
Jharkhand	7.7	67.1	22.1	77.9
Odisha	14.1	64.8	22	78
India	**30.7**	**81.4**	**46.9**	**53.1**

Source: Census of India, 2011.

The analysis of status of sanitation in eight States of North East India revealed that as per 2011 Census, Mizoram was ranked first in terms of access to sanitary latrine and Arunachal Pradesh was ranked the last. The details are presented in Table 2.5.

Table 2.5: Status of Access to Sanitation Facilities in North East India

State	Percentage of households having toilet			No Latrine (%)
	Rural	Urban	Total	
Mizoram	84.6	98.6	92	8
Manipur	85.9	95.8	89.3	10.7
Sikkim	84.1	95.2	87.2	12.8
Tripura	81.4	97.9	86	14
Nagaland	69.2	94.6	76.5	23.5
Assam	59.6	93.7	65	35
Meghalaya	51.9	95.8	62.9	37.1
Arunachal Pradesh	52.6	89.6	62	38

Source: Census of India, 2011.

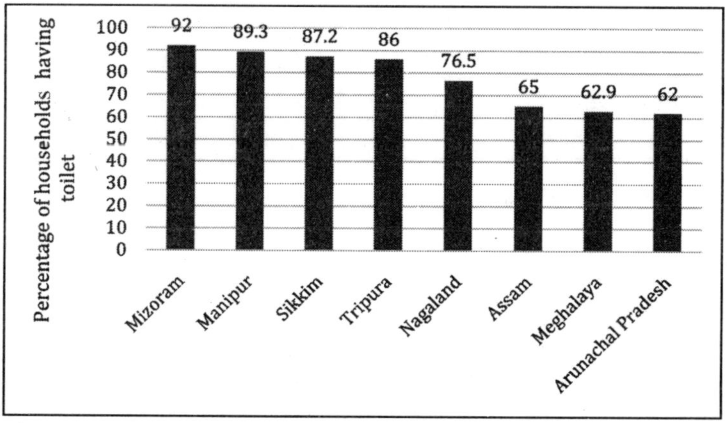

Fig. 2.2: Percentage of households having latrine facility in States of North East India (Census 2011).

In Mizoram, 92 per cent of the households had latrine facility in 2011. In the urban area of Mizoram, 98.6 per cent of the households had latrine facility and in rural area 84.6 per cent of households had latrine facility. Mizoram was followed by Manipur and Sikkim with 89.3 per cent and 87.2 per cent of households having latrine facility respectively. In Arunachal Pradesh, only 62 per cent of the households had latrine facility in 2011 (Figure 2.2). In its rural area, only 52.6 per cent of households had latrine facility, but in urban area 89.6 per cent households had latrine facility.

The analysis of status of access to type of latrine facility in Arunachal Pradesh showed that in 2001 majority of households had access to Pit latrine (25.8%). But in 2011 Census, majority had access to water closet latrine. The details are given in Table 2.6.

Table 2.6: Status of Access to Type of Latrine in Arunachal Pradesh (2001 & 2011)

Sl. No.	Type of Latrine	Percentage of households		Change (2011-2001)
		Census 2001	Census 2011	
1	Water closet	11	38.4	27.4
2	Pit latrine	25.8	18.8	-7
3	Other latrine	19.4	4.8	-14.6
A	**All type of latrine (1+2+3)**	**56.2**	**62**	**5.8**
B	**No Latrine**	**43.7**	**38**	**-5.7**
	Total (A+B)	**100**	**100**	

Source: Census of India, 2011.

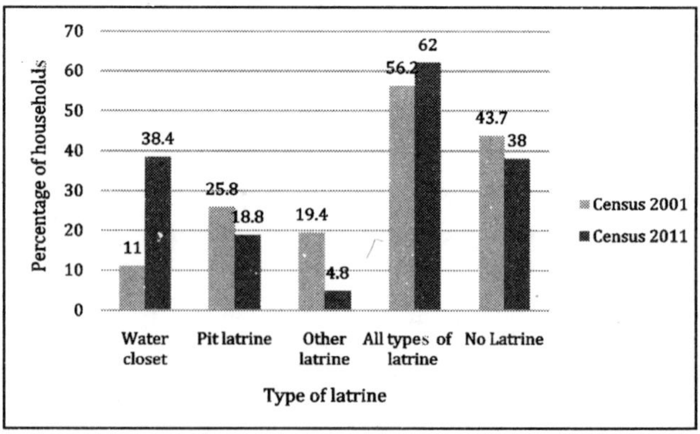

Fig. 2.3: Percentage of households having latrine facility in Arunachal Pradesh.

Table 2.6 shows that there has been significant improvement in access to proper and quality latrine in the State. In 2001 Census, only 56.2 per cent of the households had latrine facility, which improved to 62 per cent in 2011. However, it is painful that as per 2011 Census, 38 per cent of households in the State were without latrine as compared to 43.7 per cent in 2001 (Figure 2.3).

In rural Arunachal Pradesh, only 47.3 per cent of households had access to latrine in 2001 and 52.7 per cent were without latrine facility. But in urban area of the State 87 per cent of the households had latrine facility. Thus, there was a huge rural-urban gap in access to sanitation in the state. In 2011 Census, the access to latrine in rural area improved to 52.6 per cent and in urban area, it improved to 89.5 per cent. The details are given in Table 2.7 and Table 2.8.

Table 2.7: Status of Access to Sanitation Facilities in Rural Arunachal Pradesh (2001 & 2011)

Sl. No.	**Type of Latrine**	**Percentage of households**		**Change (2011-2001)**
		Census 2001	**Census 2011**	
1	Water closet	6	26.1	20.1
2	Pit latrine	24	20.4	-3.6
3	Other latrine	17.3	6.1	-11.2
A	**All types of latrine (1+2+3)**	**47.3**	**52.6**	**5.3**
B	**No Latrine**	**52.7**	**47.3**	**-5.4**
	Total (A+B)	**100**	**100**	

Source: Census of India, 2011.

The improvement in access to latrine during the period 2001 and 2011 was more in rural area than the urban area indicating declining trend in rural-urban gap in access to latrine facility. But when the quality aspect is considered, it was found that improvement was more in urban area than in rural area. In urban area of the State, during the same period the percentage of households having access to water closet latrine improved by 46.8 percentage points, but in rural area it improved by only 20.10

Table 2.8: Status of Access to Sanitation Facilities in Urban Arunachal Pradesh (2001 & 2011)

Sl. No.	Type of Latrine	Percentage of households		Change (2011-2001)
		Census 2001	Census 2011	
1	Water closet	28.1	74.9	46.8
2	Pit latrine	32.1	13.9	-18.2
3	Other latrine	26.8	0.7	-26.1
A	**All types of latrine (1+2+3)**	**87**	**89.5**	**2.5**
B	**No Latrine**	**13**	**10.5**	**-2.5**
	Total (A+B)	**100**	**100**	

Source: Census of India, 2011.

percentage points. This shows that rural area of the State suffers from both low sanitary coverage as well as poor quality of latrine.

Status of Drinking Water Supply in Arunachal Pradesh

Water is the basic necessity of life. However, majority of the people of Arunachal Pradesh were deprived of safe and treated drinking water. So, the people had to depend on natural sources like rivers, springs and lakes directly for meeting their water requirement. This posed great threat to their health. However, the coverage of drinking water supply in the State has improved over the years due to various interventions made by the government for provision of safe drinking water to the villages. But still significant proportion of people in the State lacks access to treated drinking water supply and continue to depend on untreated natural sources. As per 2011 Census, only 65.5 per cent of the households in the State had access to tap water. In rural area, only 59.3 per cent of the households had access to tap water. But in urban area, 84.2 per cent of the households had access to tap water. Thus, there is a high rural-urban disparity in access to tap water. The details are presented in Table 2.9.

Table 2.9: Distribution of households by source of drinking water in Arunachal Pradesh (Census 2011)

Source	Percentage of households		
	Rural	**Urban**	**Total**
Tap	59.3	84.2	65.5
Handpump	12.3	5.7	10.7
Tubewell	2.7	1.5	2.4
Well	5.9	5.1	5.7
Tank, Pond, Lake	1.1	0.3	0.9
River, Canal	7.6	1	5.9
Spring	7.1	1.5	5.7
Any other	4	0.7	3.2
Total	**100**	**100**	**100**

Source: Statistical Abstract of Arunachal Pradesh-2015.

Table 2.9 shows that in 2011, 15.7 per cent of the households were drawing water from the untreated natural sources (pond, lake, tank, river, canal, spring) for daily use in their households. This proportion was much higher in rural area of the State (19.8%).

The analysis of coverage of villages with drinking water revealed that in 2021 only 38.45 per cent of total villages of the State were covered with drinking water. Thus, a high proportion of villages of the State are still uncovered with drinking water which is quite disturbing fact as sanitation and good health are highly related to availability of safe water supply. The analysis of district-wise coverage of villages with drinking water showed that in 2021 only four districts, namely Dibang Valley, Siang, Upper Siang and Shi-Yomi, had more that 50 per cent of villages covered with drinking water. The details are presented in Table 2.10.

Table 2.10: District-wise number of villages fully covered by drinking water supply in Arunachal Pradesh (2021)

Sl. No.	Districts	Total Villages	Number of villages covered with drinking water	Percentage of villages covered with drinking water
1	**Dibang Valley**	**134**	**89**	**66.42**
2	Siang	105	59	56.19
3	Upper Siang	111	58	52.25
4	Shi-Yomi	113	58	51.33
5	Upper Subansiri	476	226	47.48
6	Pakke Kessang	73	32	43.84
7	Kamle	229	97	42.36
8	**Papum Pare**	**521**	**207**	**39.73**
9	West Siang	157	62	39.49
10	Changlang	332	131	39.46
11	Namsai	176	69	39.20
12	Tirap	115	45	39.13
13	Leparada	81	31	38.27
14	Lower Siang	113	43	38.05
15	Kurung Kumey	371	140	37.74
16	West Kameng	289	108	37.37
17	Lower Dibang Valley	268	99	36.94
18	Tawang	242	89	36.78
19	Lohit	116	41	35.34
20	Anjaw	306	107	34.97
21	Lower Subansiri	420	145	34.52
22	East Kameng	288	99	34.38
23	East Siang	102	32	31.37
24	Longding	76	19	25.00
25	Kra Daadi	339	49	14.45
	Total AP	**5553**	**2135**	**38.45**

Source: Statistical Abstract of Arunachal Pradesh-2021.

Dibang valley district has the highest percentage of coverage of villages with drinking water. In this district, 66.42 per cent of villages were covered with drinking water followed by Siang (56.19%), Upper Siang (52.25%) and Shi-Yomi (51.33%). In Papum Pare district, only 39.73 per cent of villages were covered with drinking water. In 2021, majority of the districts of the State (21 districts) had less than 50 per cent coverage of villages with drinking water. Among the various districts, Kra Daadi district had the lowest percentage of villages covered with drinking water followed by Longding district. In Kra Daadi and Longding districts only 14.45 per cent and 25 per cent of the villages were covered with drinking water respectively.

Thus, the policy should aim to improve the coverage of villages with drinking water as well as to reduce inter-district disparity in coverage of villages with treated drinking water.

Status of Sanitation in Papum Pare District of Arunachal Pradesh

Quality of life of the people is highly dependent on availability and access to drinking water, sanitary latrine and drainage facilities. Papum Pare district is the most urbanised district of the State with 55 per cent of the urban population. This district houses the twin capital towns of the State, namely Itanagar and Naharlagun. These two towns together accounts for 54 per cent of the total population of the district. It has a number of educational institutions of higher learning and large number of government establishments. Since majority of the people live in urban area and have access to good employment and income opportunities, the status of access to sanitation in the Circles located in proximity to the Capital is expected to be better as compared to other Circles. The most reliable data on access to sanitation in the district in the *Papum Pare District Census Handbook, 2011*. The status of sanitation in Papum Pare district has been analysed as follows:

Status of Access to Latrine Facility in Papum Pare District

Availability and access to latrine facility is the basic requirement for household sanitation and better quality of life. The analysis of data showed that in 2011, 81.15 per cent of the households in Papum Pare district had access to latrine facility and 18.86 per cent of the households had no latrine. The details are given in Table 2.11.

Table 2.11: Percentage of Households by Type of Latrine Facility in Papum Pare District of Arunachal Pradesh (Census 2011)

Circle	Total House-holds	Percentage of households having						
		Septic tank	Pit with slab	Open Pit	Public latrine	Other type*	All types of latrine	Open/ No latrine
Itanagar	13427	55.72	3.08	2.41	3.61	30.14	94.96	5.03
Naharlagun	9454	50.61	5.61	2.1	5.47	29.94	93.73	6.36
Doimukh	1689	46.36	13.5	1.72	2.61	21.08	85.27	14.74
Banderdewa	1623	36.85	5.42	8.87	4.31	13.85	69.3	30.68
Balijan	1060	10.47	4.81	34.62	2.36	12.83	65.09	34.91
Kimin	936	31.2	1.82	11.54	0	11.32	55.88	44.12
Sagalee	1135	20.53	0.09	3.52	13.39	10.75	48.28	51.72
Gumto	304	31.25	12.17	0.33	0	2.96	46.71	53.29
Toru	515	18.45	1.55	1.75	6.8	13.01	41.56	58.45
Sangdupota	534	18.16	2.43	6.18	2.81	3.74	33.32	66.67
Parang	284	5.28	1.06	2.82	2.82	21.12	33.10	66.9
Leporiang	715	9.09	0.56	0	12.03	7.83	29.51	70.49
Taraso	484	6.82	4.34	9.92	0.41	4.14	25.63	74.38
Mengio	1032	1.26	0.78	3.97	1.74	17.06	24.81	75.19
Kakoi	284	1.41	0.35	1.06	0	1.76	4.58	95.42
Papum Pare District total	**33476**	**43.92**	**4.23**	**4.04**	**4.35**	**24.61**	**81.15**	**18.86**
Rural	**13537**	**27.16**	**4.28**	**6.77**	**3.61**	**16.55**	**58.37**	**41.63**
Urban	**19939**	**55.29**	**4.19**	**2.19**	**4.85**	**30.10**	**96.62**	**3.40**

Note: *Other type includes piped sewer system, night soil disposed, service latrine.
Source: Census of India 2011—District Census Handbook Papum Pare, Arunachal Pradesh.

The analysis revealed that there was a wide rural-urban disparity in access to latrine facility in the district. While in urban area 96.62 per cent of the households had access to latrine facility, in rural area of the district only 58.37 per cent of the households had latrine facility. In urban area of the district only 3.4 per cent of the households lacked latrine facility. But in rural area of the district 41.63 per cent of the households lacked latrine and practised open defecation in 2011 which was quite alarming. Further, the analysis of quality of latrine showed that in Papum Pare district only 43.92 per cent of the households had latrine with septic tank (Table 2.11). This shows that a high proportion of the households lacked access to quality latrine facility in 2011.

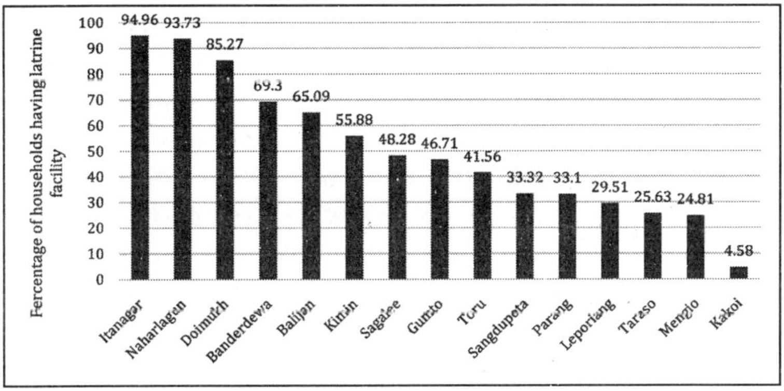

Fig. 2.4. Percentage of households having latrine facility in various Circles of Papum Pare District (Census 2011).

The analysis of status of sanitation in various Circles of the district showed that Itanagar and Naharlagun followed by Doimukh had the highest coverage of latrine facility in 2011. While Itanagar had 94.96 per cent of the households having latrine facility, Naharlagun and Doimukh had 93.73 per cent and 85.27 per cent of the households having latrine facility respectively (Figure 2.4). Among the 15 Circles of Papum Pare district in 2011, Kakoi followed by Mengio and Taraso had the lowest coverage of latrine. In Kakoi Circle only 4.58 per cent of the households had access to latrine facility in 2011. Having access to latrine does not fully ensure proper sanitation and hygiene. Proper sanitation and hygiene also depends on quality of sanitary facility. The analysis of latrine type indicated that in Itanagar and Naharlagun only 55.72 per cent and 50.61 per cent of the households have access to latrine with septic tank. This shows that though the access to latrine is high, the access to quality latrine facility is inadequate. The access to latrine with septic tank was lowest in Mengio, Kakoi and Parang Circles of the district. The analysis shows that there is a need to focus not only on the accessibility but also on the quality of facility for overall improvement in sanitation and public health.

Status of Access to Sources of Water in Papum Pare District

Water is the basic necessity of life. It is required for drinking purpose and for cleaning and washing of various things and sanitary latrines. So, regular supply of quality water is essential for improving quality of life of the people and for improving public health condition. The analysis of the status of households' access to different sources of water in Papum Pare district based on Census 2011 showed a very grim picture. It was found that only 56.91 per cent of the households had access to tap water from treated source in 2011. The details of access of households by sources of water is given in Table 2.12.

Table 2.12: Percentage of Households by Sources of Water in Papum Pare District of Arunachal Pradesh (Census 2011)

Circle	Total House-holds	Percentage of households having					
		Tap water treated	Tap water untre-ated	Cove-red well	Uncov-ered well	Hand pump & tube well	Spring/river/ pond/lake/ others
Itanagar	13427	76.32	10.82	3.96	2.09	1.58	5.23
Naharlagun	9454	65.39	11.2	7.4	5.81	2.9	7.3
Toru	515	44.85	50.68	0	0	0	4.47
Sangdupota	534	43.07	51.55	0	0	0.19	5.25
Kimin	936	37.5	15.49	10.58	12.5	0.43	23.51
Doimukh	1689	35.58	30.55	16.46	7.22	0.77	9.42

(Contd.)

Circle	Total House-holds	Percentage of households having					
		Tap water treated	Tap water untre-ated	Cove-red well	Uncov-ered well	Hand pump & tube well	Spring/river/ pond/lake/ others
Banderdewa	1623	27.66	19.16	11.58	26.37	3.57	11.66
Mengio	1032	22.19	55.62	1.16	0.58	1.45	18.99
Balijan	1060	18.11	5.09	1.23	44.72	7.92	22.92
Gumto	304	18.09	9.87	7.89	44.41	0.33	19.42
Sagalee	1135	14.01	68.55	0.7	0	0	16.73
Leporiang	715	13.15	59.86	1.68	0	0	25.32
Kakoi	284	7.75	8.8	9.86	0.35	0	73.24
Taraso	484	1.86	3.72	0.21	74.17	0	20.04
Parang	284	0.35	29.58	0	0	0	70.07
Papum Pare District	**33476**	**56.91**	**17.96**	**5.66**	**7.38**	**1.98**	**10.11**
Rural	**13537**	**31.46**	**28.87**	**5.86**	**13.34**	**2.42**	**18.05**
Urban	**19939**	**74.19**	**10.55**	**5.53**	**3.34**	**1.67**	**4.73**

Source: Census of India 2011—District Census Handbook, Papum Pare, Arunachal Pradesh.

Further, the analysis showed that 17.96 per cent of the households in the district had access to tap water from untreated sources. While 5.66 per cent of the households had access to water from covered well, 7.38 per cent of the households had access to water from uncovered well. It was found that 1.98 per cent of the households had access to water from hand pump and tube well. The analysis showed that a good proportion of the households in the district (10.11%) were relying on natural sources for water such as spring, river, canal, pond and lake in 2011. The Census 2011 data showed a high rural-urban disparity in access to quality water in the district. In rural area, only 31.46 per cent of the households had access to tap water from treated source, but in urban area 74.19 per cent of the households had access to tap water from treated source. Further, in rural area of the district, 18 per cent of the households were dependent on natural sources of water, but in urban area, only 4.73 per cent of the households were dependent on natural sources of water (Table 2.12). Thus, it can be stated that access of the rural households to quality water is much poor as compared to their urban counterparts.

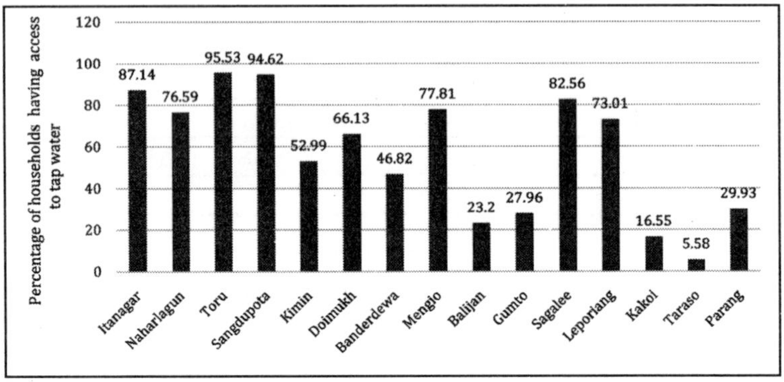

Fig. 2.5. Percentage of households having tap water in various Circles of Papum Pare District (Census 2011).

The analysis of the Circle-wise access of the households to sources of water indicated that in Toru and Sangdupota Circles highest percentage of the households had access to tap water (Figure 2.5). In Toru Circle 95.53 per cent of the households and in Sangdupota Circle 94.62 per cent of the households had access to tap water followed by ltanagar (87.14%) and Sagalee (82.56%). But in terms of access to quality water, it was found that Itanagar and Naharlagun had the highest percentage of households having access to tap water from treated source. In Itanagar and Naharlagun, 76.32 per cent and 65.39 per cent of the households had access to tap water from treated source in 2011. The analysis showed that access to tap water was the lowest in Taraso Circle and Kakoi Circle. In Taraso Circle, only 5.58 per cent of the households had access to tap water in 2011. In this Circle, 74.17 per cent of the households were dependent on uncovered well and 20 per cent of households were relying on natural sources of water like spring, river and lake in 2011. It is to be noted that in Kakoi and Parang very high proportion of households were dependent on natural sources of water. In Kakoi, 73.24 per cent of the housholds and in Parang 70 per cent of the households were dependent for water on spring, river, pond/lake.

The analysis showed that access to quality source of water in Papum Pare was inadequate in 2011. Further, there was a high rural-urban disparity in access to quality water in the district. At the same time, the study found a wide inter-Circle disparity in access to quality water in the district based on Census 2011 data.

Status of Access to Drainage Facility in Papum Pare District

Poor sanitation can increase the risk of morbidity and mortality. Availability and access to drainage facility is an important component of household sanitation. Drainage is required in each and every household for proper disposal of sewage, maintain clean environment and avoid spread of diseases. It is essential to have drainage as the lack of this facility can cause stagnation of household sewage and promote breeding of mosquitos and bacteria which are dangerous to human health. Drainage system is used to remove waste water from the households as well as to remove the excess flood or rain water. The lack of proper drainage can lead to flooding of low lying areas thereby causing damage to property and health risks. Therefore, there is a need to have an effective drainage system which can remove waste and excess water and prevent the occurrence of water logging. The drainage system can be of two types—closed drainage and open drainage. Closed drainage consists of complex network of underground pipes and is mainly used to dispose and transport sewage to a treatment plant. On the other hand, open drainage is mostly used to dispose waste water, that is not sewage. The use of open drainage for disposal of sanitary waste is extremely unhygienic and unsafe. It can be a breeding grounds for various diseases carrying pathogens.

The analysis of the Census 2011 data on the drainage system in Papum Pare district showed a gloomy state of drainage in the district. The details are presented in Table 2.13.

Table 2.13: Percentage of Households by Type of Drainage in Papum Pare District of Arunachal Pradesh (Census 2011)

Circle	Total Households	Percentage of households having			No drainage
		Closed drainage	Open drainage	All types of drainage	
Naharlagun	9454	26.37	49.97	76.34	23.67
Itanagar	13427	26.71	46.77	73.48	26.51
Kimin	936	11.32	49.79	61.11	38.89
Doimukh	1689	12.73	39.31	52.04	47.93
Banderdewa	1623	6.96	39.49	46.45	53.54
Toru	515	4.66	33.79	38.45	61.55
Balijan	1060	1.23	32.08	33.31	66.7
Sagalee	1135	7.5	24.67	32.17	67.93
Mengio	1032	1.36	14.05	15.41	84.59
Sangdupota	534	0.75	13.48	14.23	85.77
Lep[illegible]iang	715	2.1	10.49	12.59	87.41
Parang	284	4.23	7.75	11.98	88.03
Gumto	304	9.87	1.32	11.19	88.82
Taraso	484	1.86	6.61	8.47	91.53
Kakoi	284	0	1.06	1.06	98.94
Papum Pare District Total	**33476**	**20.07**	**41.59**	**61.66**	**38.34**
Rural	**13537**	**9.12**	**28.04**	**37.16**	**62.84**
Urban	**19939**	**27.5**	**50.78**	**78.28**	**21.72**

Source: Census of India 2011—District Census Handbook, Papum Pare, Arunachal Pradesh.

The analysis revealed that only 61.66 per cent of the households in the district had access to drainage facility and the rest 38.34 per cent of the households lacked drainage facility. Further, the analysis of quality of drainage indicated that that only 20.07 per cent of the households had access to closed drainage and 41.59 per cent of the households had open drainage. Moreover, there was a significant rural-urban disparity in access to drainage facility. In rural areas of the district, only 37.16 per cent of the households had access to sanitation. But in urban areas, 78.28 per cent of the households had access to drainage in 2011.

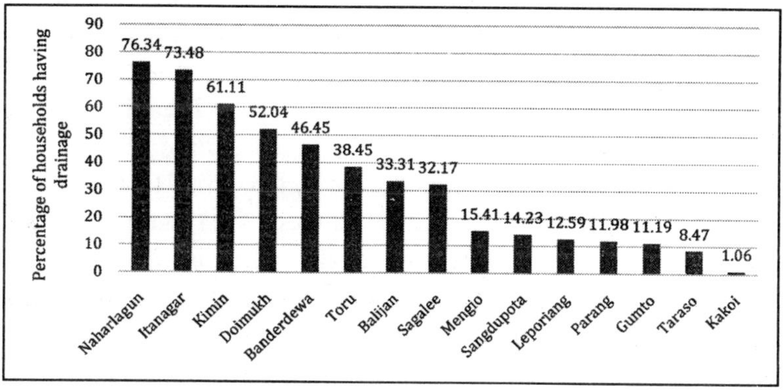

Fig. 2.6. Percentage of households having drainage in various Circles of Papum Pare District (Census 2011).

The analysis of Circle-wise status of access to drainage facility showed that Naharlagun and Itanagar followed by Kimin Circle had the highest coverage of drainage. In Naharlagun Circle, 76.34 per cent of the households had access to drainage facility and in Itanagar Circle 73.48 per cent of the households had drainage facility (Figure 2.6). However, the access of the households to closed drainage was low as most of the households had open drainage facility. The coverage of drainage was the lowest in Kakoi Circle (1.06%), Taraso Circle (8.47%) and Gumto (11.19%) Parang (11.98%). The Census 2011 data showed that the status of drainage in the district was poor and there were wide inter-Circle as well as rural-urban disparity in access to drainage facility in the district.

Conclusion

The analysis and discussion shows that sanitation in the country has improved over the years due to efforts made by the government under various plans and programmes. The access to individual household latrine improved from 36.4 per cent in 2001 Census to 46.9 per cent in 2011 Census. But still more than half of the households did not have access to the latrine. Further, the analysis showed inadequate access to quality latrine mainly in rural area. At the same time, the study found high rural-urban disparity in access to sanitary latrine. Among the States of North East India, Mizoram had the highest percentage of households with latrine facility and Arunachal Pradesh had the lowest percentage of households with latrine facility in 2011 Census. In Arunachal Pradesh, only 62 per cent of the households had latrine facility in 2011 Census. It was found that in rural Arunachal Pradesh 47.3 per cent of the households were without latrine in 2011. But in urban area of the State only 10.5 per cent of households had no latrine. This indicated huge rural-urban disparity in access to sanitation in the State. Regarding access to drinking water supply, it was found that as per 2011 Census only 65.5 per cent of the households in the State had access to tap water. In rural area, only 59.3 per cent of the households had access to tap water in 2011. Around 20 per cent of the rural households in the State were depending on river, spring, pond and lake for drinking water in 2011. The status of coverage of villages with drinking water revealed that in 2021, only 38.35 per cent of the villages in the State were covered with drinking water. Among the various districts, Dibang Valley had highest percentage of villages covered with drinking water (66.42%) followed by Siang (56.19%). Kra Daadi had the lowest percentage of the villages covered with drinking water (14.45%).

Regarding access to sanitation in Papum Pare district, it was found that in 2011 Census 81.15 per cent of the households had access to individual household latrine. But the access to proper latrine was inadequate as only 43.93 per cent had access to latrine with septic tank. Further, a high rural-urban disparity was found in access to sanitation. In rural area of the district only 58 per cent of the households had access to latrine. But in urban area, 96 per cent households had latrine facility. In rural area, access to proper sanitary latrine was low as only 27 per cent households had access to latrine with septic tank. Among the various Circles, Itanagar and Naharlagun had the highest percentage of the households having latrine facility. Kakoi, Mengio and Taraso Circles had the lowest percentage of the households having latrine facility. In Kakoi, only 4.58 per cent of the households having latrine facility. Access to drinking water is important for well-being of households. It was found that in Papum Pare district, only 56.91 per cent of households had access to tap water from treated source. In rural area of the district, the situation was worse as only 31.45 per of the households were having access to tap water from treated source in 2011 Census. But in urban area, 74 per cent of households had access to tap water from treated source indicating high rural-urban disparity. Among the various Circles, Itanagar and Naharlagun had the highest percentage of households with access to tap water from treated sources. But Parang and Kakoi Circles had the lowest percentage of households with access to tap water from treated sources. In *Parang Circle, only 0.35 per cent of households* had access to tap water from treated sources and 70 per cent of the households were relying on river, spring, lake and pond for drinking water as per 2011 Census. The analysis of data on access to drainage in Papum Pare revealed that as per 2011 Census, only 61.66 per cent households had access to drainage facility. In rural area of the district, only 37 per cent households had access to drainage facility. But in urban area, a high proportion of the households (78%) households had access to drainage facility. It indicated high rural-urban disparity in access to sanitation in Papum Pare district. Among the Circles, it was found that Naharlagun had highest percentage of households with drainage facility (76%) followed by Itanagar (73%). In Kakoi Circle, only 1.06 per cent of the total households had drainage facility in 2011 Census. The analysis showed poor access to drainage facility in rural and remote Circles of Papum Pare district.

3
Awareness of Households on Sanitation and Swachh Bharat Abhiyan

Introduction

This chapter deals with the analysis of socio-economic profile of the surveyed households and their level of awareness about sanitation and hygiene practices. The study of socio-economic characteristics of the household is important to understand their social and economic status and level of well-being. In fact, socio-economic status is an important determinants of households' access to sanitation and hygienic practices. A household with better socio-economic status is expected to have access to better sanitary facilities. The analysis of socio-economic profile of the households will give us an understanding of the level of well-being of the households. As such it can be linked to the access of the households to sanitation and hygiene. The socio-economic characteristics of a household is captured by demographic features, education, religion, economic activities, income and asset ownership pattern. These factors influence, to a great extent, the households' access to sanitation, attitude towards sanitation, sanitary and hygienic practices. Hence, there is a need to analyse the socio-economic profile of the households and understand its linkages with sanitation and sanitary practices.

Therefore, this chapter is devoted to analyse the socio-economic profile of the surveyed households and their awareness level about sanitation and Swachh Bharat Abhiyan.

Socio-Economic Profile of Households

The socio-economic profile of a household includes its demographic characteristics and economic features. The demographic characteristics of a household is captured by the variables such as age, gender, family size, religion, tribe or category, education. The economic features is given by the occupation, income, economic category, house and land ownership, ownership of other assets and durable goods. The socio-economic profile of the surveyed households has been analysed as follows:

Distribution of Household Heads by Gender

The distribution of respondent household heads by gender showed that most of them were male. The details are presented in Table 3.1.

Table 3.1: Percentage distribution of Household Heads by Gender

Gender	Frequency	Percentage
Male	573	82.4
Female	122	17.6
Total	**695**	**100**

Source: Field Survey, 2022-23.

Out of total respondents, 82.4 per cent were male and the rest 17.6 per cent were female. The gender distribution of the surveyed respondents indicates that most households in rural Arunachal are headed by male.

Distribution of Household Heads by Age Group

The examination of age distribution of household heads revealed that most of them were in the age group of 30-50 years. It was found that 60 per cent of the surveyed household heads belonged to this age group followed by 18.7 per cent in the age group of 50-60 years.

Relatively small percentage (5.5%) of the surveyed household heads was in the younger age group of 20-30 years. The distribution of surveyed household heads by age group is shown in Table 3.2.

Table 3.2: Percentage distribution of Household Heads by Age Group

Age Group	Frequency	Percentage
20-30 Years	38	5.5
30-40 Years	215	30.9
40-50 Years	204	29.4
50-60 Years	130	18.7
60 years and above	108	15.5
Total	**695**	**100**

Source: Field Survey, 2022-23.

Further, it was found that 15.5 per cent of the household heads were in the age group of 60 years and above. The mean age of the household heads was 45 years with standard deviation of 11.55. While the minimum age of household head was 22 years, the maximum age was 80 years.

Distribution of Household Heads by Education Level

Education is an important factor influencing sanitation and sanitary practices. An examination of education level of the surveyed household heads revealed high incidence of illiteracy among the households. It was found that 26 per cent of the household heads were illiterate. Thus, lack of education could be a factor acting as a barrier for access to better sanitation. While 3.9 per cent of the household heads were educated up to primary level, 11.7 per cent of them were educated up to upper primary level. The details on education level of the surveyed household heads is given in Table 3.3.

Table 3.3 shows that 20.9 per cent of household heads were educated up to secondary level and 14.8 per cent were educated up to higher secondary level. A good number of household heads

(16%) were educated up to graduation level and only 4 per cent of the household heads had education of post-graduation level.

Table 3.3: Percentage distribution of Household Heads by Education Level

Education Level	Frequency	Percentage
Illiterate	181	26
Primary level	27	3.9
Upper primary	81	11.7
Secondary	145	20.9
Higher Secondary	103	14.8
Graduate	111	16
Post Graduate	28	4
Others	19	2.7
Total	**695**	**100**

Source: Field Survey, 2022-23.

Distribution of Household Heads by Primary Occupation

Occupation is an important determinants of a household's well-being and access to basic facility. The analysis of primary occupation the surveyed household heads showed that they were engaged in diverse livelihood activities. It was found that around 40 per cent of them were engaged in crop farming and 24.8 per cent were doing business as their primary occupation. The details of primary occupation of the households' heads are presented in Table 3.4.

Table 3.4: Percentage distribution of Household Heads by Primary Occupation

Primary occupation	Frequency	Percentage
Crop farming	277	39.9
Business	172	24.8
Govt. service	127	18.3
Private service	40	5.8
Others (self-employed)	32	4.6

(*Contd.*)

Primary occupation	Frequency	Percentage
Plantation crop farming	19	2.7
Wage labour	8	1
Livestock rearing	7	1
Craft Making	7	1
Forest based activities	4	0.6
Sericulture	2	0.3
Total	**695**	**100**

Source: Field Survey, 2022-23.

Table 3.4 shows that primary occupation of 18.3 per cent of the heads of the households was government service and 2.7 per cent of them were doing plantation crop farming as primary occupation. It was found that only a few respondents were engaged as a wage labourer, livestock rearing, craft making and forest based-activities.

Distribution of Households by Annual Household Income Group

Income is an important indicator of household's well-being and its capability to access sanitary and other facilities. A household can generate income from various sources. Total income of the household is the sum total of the incomes earned by various members from different sources. Hence, an attempt was made to examine the annual income of the surveyed households. It was found that most of the households had income below Rs. 5 lakh. The details are presented in Table 3.5.

Table 3.5: Percentage distribution of Households by Annual Household Income Group

Income Group	Frequency	Percentage
Below 2.5 Lakh	89	12.81
2.5-5 Lakh	247	35.54
5-10 Lakh	292	42.01
10-15 Lakh	64	9.21
15 Lakh & above	3	0.43
Total	**695**	**100**

Source: Field Survey, 2022-23.

It was found that most of the surveyed households were in the income group of 5-10 lakh followed by 2.5-5 lakh group. While 42 per cent of the households were in the income group of 5-10 lakh, 35.54 per cent of them were in the income group of 2.5-5 lakh. However, a good numbers of households (12.81%) belonged to lower income group of below 2.5 lakh. Since income is an important factor influencing access to sanitation, there is need to create alternative sources of income for the people in rural areas for improving sanitation among the masses.

Distribution of Households by Religion

Region and faith also influence human behaviour and attitude towards various social and economic activities in a society. In fact, it teaches human beings about the values, ethics, norms and good practices for living a healthy life. It can also influence people's attitude towards sanitation and sanitary practices. The distribution of surveyed households by religion is presented in Table 3.6.

Table 3.6: Percentage distribution of Household Heads by Religion

Religion	Frequency	Percentage
Hindu	9	1.3
Muslim	1	0.1
Buddhist	1	0.1
Christian	612	88.1
Donyi Polo	67	9.7
Others	5	0.7
Total	**695**	**100**

Source: Field Survey, 2022-23.

It was found that most of the surveyed households (88%) belonged to Christians followed by Donyi Polo (9.7%) and Hindu (1.3%). Only a small proportion of the surveyed households belonged to Buddhist and Muslim religions. Traditionally, the entire population belonging to *Tani* group were professing the indigenous faith 'Donyi Polo'. However, in the last few decades, a large number of people have adopted Christianity. Thus, it can be inferred that the habits, attitudes, sanitation culture and practices of the people have been influenced by the changed in religion.

Distribution of Households by Tribe

The studied district, being the district having State capital, is inhabited by a large number of tribal groups. In fact, it inhabited by all of the tribes of the State. Each tribal group of the State has its own custom, tradition and social as well as economic practices. The custom, tradition and culture are important factors influencing the people's life style, attitudes and practices. Thus, there could be inter-tribe differences in attitude and access towards sanitation and hygiene. The analysis of tribe-wise distribution of surveyed households showed that most of them belonged to Nyishi tribe. The details are presented in Table 3.7.

Table 3.7: Percentage distribution of Household Heads by Tribes

Tribe	Frequency	Percentage
Nyishi	557	80.2
Nocte	1	0.1
Adi	2	0.3
Galo	13	1.9
Apatani	5	0.7
Tagin	14	2
Hindu	2	0.3
Others	101	14.5
Total	**695**	**100**

Source: Field Survey, 2022-23.

It was found that 80.2 per cent of the households belonged to Nyishi tribe. This could be due to the fact that rural areas of Papum Pare district is mainly inhabited by Nyishi community and the study was based on field work in rural areas of this district. However, 2 per cent of the surveyed were from Tagin tribe and 1.9

per cent were from the Galo tribe. It was found that 14.5 per cent of the households belonged to other tribal groups.

Distribution of Households by Economic Category

Economic status is an important indicator of a household's level of well-being. In India, households are categorized into below poverty line (BPL) and above poverty line (APL) using the criterion of certain minimum level of per capita consumption expenditure (Rs. 1286 in urban areas and 1059.42 in rural areas). The study used BPL card holding status to categorise households into BPL and APL. Those households who were in possession of BPL card were considered as BPL and others were considered as APL. The details are presented in Table 3.8.

Table 3.8: Percentage distribution of Households by Economic category

Response	Frequency	Percentage
BPL	417	60.3
APL	276	39.7
Total	**695**	**100**

Source: Field Survey, 2022-23.

It was found that 60.3 per cent of the households belonged to BPL and the remaining 39.7 per cent of them belonged to APL. The high proportion of BPL does not necessarily mean high incidence of poverty in the district. It could be due to inclusion error, that is, inclusion of non-BPL households in BPL category by issuing of BPL cards.

Distribution of Household by Family Size

Family size is an important demographic factor which influences household's choice pattern and access to basic facilities. In fact, it determines the household's per capita income which is an important indicator of standard of living. The popular saying 'small family is a happy family' is highly relevant as far as access of household to basic needs and sanitation is concerned. The analysis showed that most of the surveyed households had family size of 4-6 members (semi-medium size). The details are presented in Table 3.9.

Table 3.9: Percentage distribution of Household by Family Size

Family Size	Frequency	Percentage
1 to 3	106	15.25
4 to 6	402	57.84
7 to 9	134	19.28
10 and above	53	7.63
Total	**695**	**100**

Source: Field Survey, 2022-23.

Table 3.9 shows that 57.84 per cent of the surveyed households had a family size of 4-6 members (semi-medium size) and 19.28 per cent of them had family size of 7-9 members (medium size). Interestingly, a good proportion of the surveyed households (15.25%) belonged to small size family (1-3 members). It is to be noted that only 7.63 per cent of the households had large family size (10 members and above). The analysis shows that there has been change in demographic structure of tribal households with change in time and modernization. Traditionally, tribal community used to live in a joint family with more than 10 members living together and eating food from the same kitchen. The change in family size towards small and medium is an indication of progress of tribal society.

Distribution of Households by Number of Earning Members

Household's income is contributed by different members. Some households have just one earning member, while others have more than one earning members. More is the number of earning members, higher is the household income and better is the standard of living. The analysis showed that half of the surveyed households had only one earning member. This could be an explanation for low standard of living and lack of access of tribal community to basic household facilities including sanitation. The details are presented in Table 3.10.

Table 3.10: Percentage distribution of Households by Number of Earning Members

Earning members	Frequency	Percentage
Only One	349	50.2
Two to Three	310	44.6
More than 3	36	5.2
Total	**695**	**100**

Source: Field Survey, 2022-23.

Table 3.10 shows that 50.2 per cent of the households had only one earning member. However, 44.6 per cent of the households had two-three earning member. It was found that only 5.2 per cent of the surveyed households had more than three earning members.

Distribution of Households by House Ownership

Housing is an important component of human life. Therefore, access to proper housing is essential for good health and hygienic life of the people. The analysis of surveyed households by house ownership showed that most of them had their own house. The details are presented in Table 3.11.

Table 3.11: Percentage distribution of Households by House Ownership

House Type	Frequency	Percentage
Own house	625	89.9
Rented house	47	6.8
Quarter	18	2.6
Others	5	0.7
Total	**695**	**100**

Source: Field Survey, 2022-23.

It was found that 89.9 per cent of the surveyed households were living in the house owned by them which itself is a good indication of housing among the tribal community. But 6.8 per cent of the surveyed households were living in rented house and 2.6 per cent of them were living in quarters (house provided by employers). Others category include living in a rent free house of relatives.

Distribution of Households by Asset Holdings

Asset ownership is an important indicator of standard of living of the households as well as its income. The most important asset that a tribal household own is land. The asset ownership pattern of the surveyed households is presented in Table 3.12.

Table 3.12: Percentage distribution of Households by Asset Holdings

Assets	Number of HH	Percentage of HH
Radio	9	1.3
TV	470	87.2
Refrigerator	485	60.3
Washing Machine	230	33
Motor Cycle	331	47.7
Car	282	40.6
Commercial Vehicle	23	3.3
Land	512	73.7

Source: Field Survey, 2022-23.

The analysis shows that 73.7 per cent of the surveyed households had ownership of land which clearly reflects their capability to sustain their livelihood through agricultural practices and other allied activities like livestock rearing and fisheries. While 87.2 per cent of the surveyed households had television, 60.3 per cent of them also had refrigerator and 33 per cent had washing machine too. Regarding ownership of two-wheeler and vehicle, it was found that 47.7 per cent of the surveyed households had motorcycle and 40.6 per cent had car. Further, 3.3 per cent of the surveyed households were found to possess commercial vehicle. This shows the emergence of entrepreneurs among the tribal community in rural area of the State which is really an encouraging development.

Awareness Level of Households on Sanitation and Hygiene

Awareness about the importance of sanitation and hygiene is a very important factor determining a household's access to sanitation facilities and sanitary habits. A household with members having very good level of awareness about the importance of sanitation and hygiene is expected to have access to better sanitation and sanitary practices. Hence, it was considered important to examine the level of awareness of the surveyed households about the various aspects of sanitation. The level of awareness of the households about various aspects of sanitation has been measured using a five-point Likert scale (1=poor, 2=average, 3=good, 4=very good, 5= excellent).

The levels of awareness of the surveyed households heads on various aspects of sanitation are discussed as follows:

Awareness on Poor Sanitation and Communicable Diseases

Sanitation is important for maintaining good public health. The poor sanitation can cause a number of communicable diseases. Therefore, people need to have high level of awareness about the ill effects of poor sanitation. Regarding the level of awareness of surveyed respondents on poor sanitation and communicable diseases, it was found that most of the respondents had good level of awareness followed by very good and average. The details are presented in Table 3.13.

Table 3.13 shows that 40.9 per cent of the surveyed households had good and 32 per cent of them had very good level of awareness on the link between poor sanitation and communicable diseases. Hoverer, 18 per cent of the respondents had average and 1.9 per cent of them had poor level of awareness on this aspect which is a matter of concern. Further, it was found that 7.2 per cent of the respondents had excellent level of awareness on the link between poor sanitation and communicable diseases.

Table 3.13: Poor sanitation leads to communicable diseases

Awareness Level	No. of Respondents	Percentage
Poor	13	1.9
Average	125	17.9
Good	284	40.9
Very Good	223	32.1
Excellent	50	7.2
Total	**695**	**100**

Source: Field Survey, 2022-23.

Awareness of Households on Importance of Sanitation and Diseases

Proper sanitation is the key to good health of the people. In fact, access to safe water and sanitation to the community is basic precondition for improving health and alleviating poverty (Nath, 2017). However, access to better sanitation and hygiene, among other factors, depends on the level of awareness of the people. The analysis of the level of awareness of the respondents on proper sanitation and good health indicated that most of the households had good and very good level of awareness on this aspect. The details are presented in Table 3.14.

Table 3.14: Proper sanitation keeps us healthy and free from diseases

Awareness Level	No. of Respondents	Percentage
Poor	14	2
Average	123	17.7
Good	294	42.3
Very Good	227	32.7
Excellent	37	5.3
Total	**695**	**100**

Source: Field Survey, 2022-23.

It was found that while 42.3 per cent of the respondents had good level of awareness, 32.7 per cent of them had very good level of awareness about the link between proper sanitation and good health. At the same time, 5.3 per cent of the respondents had excellent level of awareness on the aspects. However, it is to be noted that 17.7 per cent of the respondents had average level of awareness and 2 per cent of them had poor level of awareness about the proper sanitation and health.

Awareness on Clean Drinking Water and Water Borne Diseases

Households' access to clean drinking water is highly essential to ensure good health of the members and stay free from water borne diseases. It is, therefore, utmost important to achieve universal access to all households to safe and clean drinking water. The access of a household to clean drinking water, to a great extent, depends on the awareness level of the head of the household about the essence of clean drinking water in addition to other factors. The analysis of the awareness level of the respondents about clean drinking water prevents water borne diseases showed that a good proportion of the respondents had good and very good level of awareness. The details are presented in Table 3.15.

Table 3.15: Clean drinking water prevents water borne diseases

Awareness Level	No. of Respondents	Percentage
Poor	12	1.7
Average	123	17.8
Good	291	42.2
Very Good	208	30.2
Excellent	56	8.1
Total	**690**	**100**

Source: Field Survey, 2022-23.

Table 3.15 shows that 42.2 per cent of the respondents had good level of awareness and 30.2 per cent of them had very good level of awareness about the aspect of clean drinking water prevents water borne diseases. Further, it was found that 8 per cent of the respondents had excellent level of awareness regarding the importance of clean drinking water for preventing diseases. On the other hand, it was found that nearly 18 per cent of the respondents had average and 1.7 per cent had poor level of awareness about importance of clean drinking water for preventing diseases.

Awareness on Proper Drainage for Clean Household Surroundings

Access to proper drainage and sewage system is quite essential for disposal of sewage and for keeping the household surrounding clean. Among other factors, level of awareness of household head could be an important determinants of household's access to proper drainage system. Hence, an attempt was made to examine the level of awareness of the respondent heads of the surveyed households about the importance of proper drainage for clean household surroundings. The analysis indicated that nearly half of the surveyed heads of the households had good level of awareness about the importance of proper drainage for clean household surroundings. The details are presented in Table 3.16.

Table 3.16: Proper drainage needed for clean household surroundings

Awareness Level	No. of Respondents	Percentage
Poor	12	1.7
Average	135	19.6
Good	308	44.6
Very Good	199	28.9
Excellent	36	5.2
Total	**690**	**100**

Source: Field Survey, 2022-23.

Table 3.16 shows that 44.6 per cent of the respondents had good level of awareness and 28.9 per cent of them had very good level of awareness about the importance of proper drainage for clean household surroundings. Further, it was found that 5.2 per cent of the respondents had excellent level of awareness regarding the importance of proper drainage for clean household surroundings. On the other hand, it was found that 19.6 per cent of the respondents had average and 1.7 per cent had poor level of awareness about importance of proper drainage for clean household surroundings.

Awareness on Importance of Clean Toilet and Disposal of Human Waste

An important component of proper sanitation is the proper disposal of human excreta. Proper disposal of human faeces is utmost important as human contact to it can cause a number of infectious diseases such as diarrhoea, typhoid, ascariasis etc. It is important to dispose human waste to avoid water pollution and minimise the possibility of spreading of disease. Hence, the study made an attempt to examine the level of awareness of surveyed household heads about the importance of clean toilet and disposal of human waste. The details are presented in Table 3.17.

Table 3.17: Clean toilet and proper disposal of human waste prevents spread of diseases

Awareness Level	No. of Respondents	Percentage
Poor	12	1.7
Average	119	17.3
Good	310	44.9
Very Good	212	30.7
Excellent	37	5.4
Total	**690**	**100**

Source: Field Survey, 2022-23.

Table 3.17 shows that a fairly good proportion of the respondents (44.9%) had good level of awareness and 30.7 per cent of them had very good level of awareness about the importance of clean toilet and disposal of human waste. Further, it was found that 5.4 per cent of the respondents had excellent level of awareness about this aspect of sanitation. On the other hand, it was found that 17.3 per cent of the respondents had average and 1.7 per cent had poor level of awareness about the importance of clean toilet and disposal of human waste for preventing spread of diseases. This shows that there is need for conducting awareness programme to sensitise people about the importance of maintaining clean toilet and disposal of human waste.

Awareness about Importance of Filtered and Boiled Water

Consumption of safe drinking water is highly essential for maintaining good health of the community. Therefore, it is extremely essential to ensure access of all households to safe drinking water. For maintaining good health of family members, a household is required to ensure access to properly filtered or boiled water. However, the access of a household to safe drinking water is determined by a number of factors. The level of awareness of the household's head could be an important factor determining access to safe drinking water. Hence, an attempt was made to analyse the level of awareness of the respondents about the importance of filtered and boiled water for maintaining good health. The details of awareness level of the respondents on this aspect of sanitation is presented in Table 3.18.

Table 3.18: Filtered/boiled water prevents spread of diseases

Awareness Level	No. of Respondents	Percentage
Poor	12	1.7
Average	126	18.3
Good	289	41.9
Very Good	209	30.3
Excellent	54	7.8
Total	**690**	**100**

Source: Field Survey, 2022-23.

Table 3.18 shows that a fairly good proportion of the respondents (41.9%) had good level of awareness and 30.3 per cent of them had very good level of awareness about the importance of filtered and boiled water for preventing spread of diseases. Further, it was found that 7.8 per cent of the respondents had excellent level of awareness about this aspect of sanitation. On the other hand, it was found that 18.3 per cent of the respondents had average and 1.7 per cent had poor level of awareness about the importance of safe drinking water for preventing spread of diseases. This shows that there is need to sensitise people about the importance of safe drinking water for preventing the spread of diseases.

Awareness about Importance of Use of Soap

Human surroundings is full of infectious micro-organisms like bacteria and viruses. In the course of carrying out various activities, human being is always at a high risk of getting infected by the pathogens. Hence, it is highly essential for the people to use soap and other disinfecting items to keep themselves safe and prevent the spread of diseases. But the use of soap depends on level of awareness among other factors. The analysis of the level of awareness of the surveyed household heads about the importance of use of soap showed that three-fourth of them had good and very good level of awareness. The details are presented in Table 3.19.

Table 3.19: Use of soap prevents spread of diseases

Awareness Level	No. of Respondents	Percentage
Poor	13	1.9
Average	123	17.8
Good	308	44.6
Very Good	206	29.9
Excellent	40	5.8
Total	**690**	**100**

Source: Field Survey, 2022-23.

Table 3.19 shows that 44.6 per cent of the respondents had good level of awareness and 29.9 per cent of them had very good level of awareness about the importance of use of soap for preventing spread of diseases. Further, it was found that 5.8 per cent of the respondents had excellent level of awareness about the importance of use of soap. On the other hand, it was found that 17.8 per cent of the respondents had average level of awareness about use of soap for preventing spread of diseases. It was found that only 1.9 per cent of the respondents had poor level of awareness about the importance of use of soap. This shows that awareness level of households is good as far as use of soap is concerned. However, there is a scope for sensitizing the people about the importance of use of soap and sanitiser for preventing the spread of diseases.

Awareness about Importance of Segregating of Waste

Household waste is one of the important sources of waste. The daily household activities of consumption generates waste which needs to be segregated for better management. Waste can be broadly grouped into two categories organic and inorganic wastes. Organic wastes such as remains of vegetables and food are biodegradable and can be easily decomposed by the natural processes. However, the inorganic wastes such as plastics and chemicals are non-biodegradable and cannot be assimilated by the natural processes. Hence, there is need to segregate household's waste into organic and inorganic for better waste management. The study made an attempt to examine the level of awareness of household heads about the importance of waste segregation. The analysis showed that most of the respondents had good level of awareness about the importance of waste segregation for better waste management.

Table 3.20: Segregating of waste into organic and inorganic for better waste management

Awareness Level	No. of Respondents	Percentage
Poor	14	2
Average	143	20.8
Good	294	42.6
Very Good	194	28.1
Excellent	45	6.5
Total	**690**	**100**

Source: Field Survey, 2022-23.

Table 3.20 shows that 42.6 per cent of the respondents had good level of awareness and 28.1 per cent of them had very good level of awareness about the importance of waste segregation for better waste management. However, it was found that 20.8 per cent of the respondents had average level of awareness about the importance of waste segregation. This calls for attention of the policy makers for necessary intervention. The study found that 6.5 per cent of the respondents had excellent level of awareness about the importance of waste segregation. It is encouraging to note that only a few respondents (2%) had poor level of awareness about the importance of waste segregation into organic and inorganic.

Awareness about Importance of Proper Disposal of Waste

Proper disposal of solid as well as liquid waste is pivotal for keeping the household's surrounding clean and prevent the spread of diseases. The analysis of the level of awareness of the respondents showed that most of them had good and very good level of awareness on this aspects which is quite encouraging for sanitation and hygiene. The details are presented in Table 3.21.

Table 3.21: Proper disposal of waste prevents spread of diseases

Awareness Level	No. of Respondents	Percentage
Poor	15	2.2
Average	124	17.9
Good	313	45.4
Very Good	207	30
Excellent	31	4.5
Total	**690**	**100**

Source: Field Survey, 2022-23.

Table 3.21 shows that 45.4 per cent of the respondents had good level of awareness and 30 per cent of them had very good level of awareness about the importance of proper disposal of waste for preventing spread of diseases. However, only 6.5 per cent of the respondents were found to have excellent level of awareness about the importance of proper disposal of waste. Further, it was found that 17.9 per cent of the respondents had average and 2.2 per cent of them had poor level of awareness about the importance of waste segregation. Hence, there is a need to sensitise people on proper waste disposal for better sanitation.

Awareness about Open Defecation and its Effect

One of themain aims of Swachh Bharat Mission is to eliminate open defecation in the country. Elimination of open defecation is important to avoid human contact to faeces and prevent spread of diseases. The achievement of open defecation free society to a great extent depends on awareness level of the people about its effects. Hence, an attempt was made to examine the level of awareness of the respondents of surveyed households about the effects of open defecation on public health. The analysis showed that around one-third of the households had very good level of awareness about the effects of open defecation. The details are presented in Table 3.22.

Table 3.22: Open defecation can cause spread of disease

Awareness Level	No. of Respondents	Percentage
Poor	12	1.7
Average	128	18.6
Good	262	38
Very Good	212	30.7
Excellent	76	11
Total	**690**	**100**

Source: Field Survey, 2022-23.

Table 3.22 shows that 38 per cent of the respondents had good level of awareness and 30.7 per cent of them had very good level of awareness about the effects of open defecation on public health. Further, 11 per cent of the respondents were found to have excellent level of awareness about this aspect of sanitation. However, it is to be pointed out that 18.6 per cent of the respondents had average and 1.7 per cent of them had poor level of awareness about the effects of open defecation on public health. Hence, there is a need to sensitise people about the harmful effects of open defecation for better sanitation.

Descriptive Statistics of Awareness Level of Surveyed Households

The awareness level of the surveyed households' heads on various aspects of sanitation has been measured using a five point Likert Scale. The score 1 indicates poor level of awareness and 5 indicates excellent level of awareness of the households' heads. The descriptive statistics of the level of awareness of the surveyed households' heads on various aspects of sanitation are presented in Table 3.23.

Table 3.23: Descriptive Statistics of Awareness Level of Surveyed Households on Various aspects of Sanitation and Hygiene
(1=poor, 2=average, 3=good, 4=very good, 5=excellent)

Aspects of Sanitation	No. of Respondents	Min.	Max.	Mean score	Std. Deviation
Poor sanitation leads to communicable diseases	695	1	5	3.25	0.901
Proper sanitation keeps us healthy and free from diseases	695	1	5	3.21	0.871
Clean drinking water prevents water borne diseases	690	1	5	3.25	0.906
Proper drainage needed for clean household	690	1	5	3.16	0.863
Clean toilet prevents spread of diseases	690	1	5	3.21	0.854
Filtered/boiled water prevents spread of diseases	690	1	5	3.24	0.905
Clean drainage prevents diseases	690	1	5	3.17	0.841
Use of soap prevents diseases	690	1	5	3.20	0.868
Segregating of waste into organic and inorganic	690	1	5	3.16	0.901
Proper disposal of waste prevents spread of diseases	689	1	5	3.17	0.849
Open defecation can cause spread of diseasess	690	1	5	3.31	0.958

Source: Field Survey, 2022-23.

Table 3.23 shows that the mean score of awareness level of heads of the surveyed households on all the aspects of sanitation under consideration ranged between 3.16 to 3.31. This indicated that on an average the respondents had good level of awareness about the various aspects of sanitation. The standard deviation of the scores were low which indicated unanimity or uniformity of awareness level of the respondents. This shows that there is a scope to improve the level of awareness of rural households on various aspects of sanitation through appropriate interventions. Among the various means of sanitation, the mean score was found to be the highest for effect of open defecation, impact of poor sanitation and importance of clean drinking water. The mean score was found to be the lowest for importance of drainage and waste segregation. It implies that households need to be sensitised on importance of having proper drainage and segregation of waste for achieving the goals of Swachh Bharat Mission.

Awareness of Households about Swachh Bharat Abhiyan

Swachh Bharat Abhiyan is so far as the country's biggest cleanliness drive. The programme Swachh Bharat Abhiyan (SBG) was lunched by the Government of India as a nation-wide programme on 2nd October, 2014, the birth anniversary of the father of nation Mahatma Gandhi who was the champion of sanitation movement in the country during the colonial period. It is different from the earlier programmes as it envisages to make sanitation a mass movement by encouraging active participation community in the cleanliness drive. The mission Swachh Bharat Abhiyan aims to eradicate open defecation in the country by 2019 with a plan to construct 12 crore toilets in rural India. Open defecation had been a rampant in the country as per 2011 Census report and it had been a major source of surface water pollution and threat to public health. According to United Nations, India has the largest number of people defecating in open and it is the home to largest number of slums in Asia (Singh, 2019). Poor sanitation was not only due to poverty and low income, but also due to mindset, cultural barrier and superstition among the rural population. The mission, therefore, aims to bring about a behavioural change among the people by promoting cleanliness and hygiene; to improve sanitation coverage by providing financial support for constructing individual household latrine (IHHL) and community sanitary complexes and separate toilets for girls in the schools; to develop community managed sanitation system focusing on scientific solid and liquid waste management systems for overall cleanliness of rural areas and to motivate communities and Panchayati Raj Institutions to adopt sustainable sanitation practices and facilities through awareness creation and health education.

The progress of the mission indicates that during the period 2nd October, 2014 to 2nd October, 2022, 10.9 crore toilets have been built in the country under the mission. As per the report of National Annual Rural Survey 2019-20, 99.6 per cent of the rural households had toilet facility and availability of water and 95.2 per cent households were using the facility (Govt. of India, 2022). In Arunachal Pradesh, 1.45 lakh toilets were built under the Swachh Bharat Mission during 2014-2022.

Thus, Swachh Bharat Abhiyan is a massive programme of the country which seeks to achieve open defecation free India and improve the overall quality of life of the people through various forms of intervention, such as financial incentive, motivation, creation of awareness, building up of facilities for waste management and involvement of communities and local institutions. Under the mission, a financial incentive of Rs. 12,000 is provided to eligible beneficiaries to construct individual household latrine and water storage facility for hand washing. However, the overall success of the programme depends on the awareness level of the people, particularly in rural areas and their involvement and participation in the programme as well as its effective implementation by the government machineries. Hence, an attempt was made to examine the level of awareness among the surveyed households about the mission and their participation and involvement in the programme.

The awareness of the surveyed households about the Swachh Bharat Abhiyan is discussed as follows:

Participation in Swachh Bharat Abhiyan Activities

The attitude and willingness of the people are important for successful implementation of any programme. The analysis showed that majority of the surveyed respondents were of the opinion that everyone should participate in the Swachh Bharat Abhiyan activities. The details are presented in Table 3.24.

Table 3.24: Everyone should participate in Swachh Bharat Abhiyan

Response/Attitude	No. of Respondents	Percentage
Yes	609	87.6
No	67	9.7
Don't Know	19	2.7
Total	**695**	**100**

Source: Field Survey, 2022-23.

Table 3.24 shows that 87.6 per cent of the surveyed respondents were of the view that everyone should participate in the Swachh Bharat Abhiyan activities. This implies that people are well motivated and have positive attitude towards the mission which is very encouraging for the sanitation programme. However, some respondents (9.7%) opined that there is no need for everyone to participate in the programme and 2.7 per cent stated don't know.

Awareness about Swachh Bharat Abhiyan and Cleanliness/ Hygiene

The basic aim of the Swachh Bharat Abhiyan is to bring about a improvement in quality of life of the people in rural areas by promoting cleanliness and hygiene. The analysis of the data on awareness level of the surveyed households about this aim of the mission revealed that a high percentage of the respondents had awareness about this aim of the mission (Table 3.25).

Table 3.25: Swachh Bharat Abhiyan aims to promote cleanliness and hygiene

Response/Attitude	No. of Respondents	Percentage
Yes	680	97.90
No	15	2.10
Total	**695**	**100**

Source: Field Survey, 2022-23.

It was found that 97.90 per cent of the respondents had awareness about the aim of the mission to promote cleanliness and hygiene in the communities. However, 2.10 per cent of the respondents were not aware about it.

Awareness about Swachh Bharat Abhiyan and Technical and Financial Support

One of the most important aspects of the Swachh Bharat Abhiyan is that it provides for technical and financial support to the households for constructing individual household latrine (IHHL). The availing of benefits by the households from the programme depends on their awareness about the provision. Hence, an attempt was made to examine the awareness of the surveyed households about this provision of the mission. The analysis showed that most of the respondents were aware about this provision of the programme. The details are presented in Table 3.26.

Table 3.26: Provision of technical and financial support for construction of IHHL under Swachh Bharat Abhiyan

Response/Attitude	No. of Respondents	Percentage
Yes	660	95
No	11	1.6
Not responded	24	3.4
Total	**695**	**100**

Source: Field Survey, 2022-23.

It was found that 95 per cent of the respondents were aware about the mission's provision of technical and financial support for constructing individual household latrine (IHHL). However, it was found that 1.6 per cent of the respondents were not aware about this provision of the mission.

Awareness about Swachh Bharat Abhiyan and Community Sanitary Complexes

The Swachh Bharat Abhiyan aims to improve the quality of life and overall cleanliness of villages. Therefore, it provides for technical and financial support for construction of community sanitary complexes with proper structure, latrine and water supply. The analysis showed that most of the respondents were aware about this provision of the mission. The details are presented in Table 3.27.

Table 3.27: Provision of community sanitary complexes under Swachh Bharat Abhiyan

Response/Attitude	No. of Respondents	Percentage
Yes	234	33.7
No	35	5
Not responded	426	61.3
Total	**695**	**100**

Source: Field Survey, 2022-23.

Table 3.27 shows that 33.7 per cent of the respondents were aware about the provision of technical and financial support for constructing community sanitary complexes in their villages. However, it is to be noted that 5 per cent of the respondents did not know about this provision of the mission.

Awareness about Swachh Bharat Abhiyan and Waste Management

Waste management is one of the primary requirements for improving sanitation and hygiene of the communities. The mission's objective is also to promote community managed sanitation system focusing on scientific solid and liquid waste management systems for overall cleanliness of rural areas. To ensure better waste management, households are given SBA dustbin and they are encouraged to dispose waste in community bin or municipal bin. The wastes are to be collected by the vehicles of Urban Local Bodies (ULBs) and carried away to the treatment plants or disposed-of in the designated landfills. The analysis of the awareness of the people on better waste management practices under the Swachh Bharat Abhiyan showed that most of them were aware about this provision of the mission.

Table 3.28: Swachh Bharat Abhiyan aims to improve cleanliness in villages through better waste management practices

Response/Attitude	No. of Respondents	Percentage
Yes	635	91.32
No	60	8.68
Total	**695**	**100**

Source: Field Survey, 2022-23.

Table 3.28 shows that 91.32 per cent of the respondents were aware about the provision of the mission for better waste management practices to improve overall cleanliness of the villages. But it was found that 8.68 per cent of the respondents were not aware about this provision of the mission.

Swachh Bharat Abhiyan and Incentive to Households

The mission Swachh Bharat Abhiyan seeks to improve sanitation coverage in the country by promoting construction of individual household latrine (IHHL). In a developing country like India, large number of people still belong to below poverty line group. Though the proportion of people below poverty line has gone down over the years from 37 per cent in 2004-05 to 22 per cent in 2011-12 (Planning Commission, 2012). The percentage of people below poverty line has further declined to 16.4 per cent in 2020 (UNDP, 2022). But in absolute term still millions people live in poverty without access to basic needs of life and sanitary facilities. In fact, India has the largest number of poor people worldwide (228.9 million in 2020) (UNDP, 2022). Hence, the goal of improved sanitation could be achieved in the country through the provision of financial support to the poor households for constructing individual household latrine (IHHL). Under the Swachh Bharat Abhiyan there is a provision for financial incentive of Rs. 12,000 per unit to BPL households for constructing the latrine and water storage facility. Hence, an attempt was made to examine the proportion of households who have received financial support or subsidy for constructing the latrine facility under the mission. The details are presented in Table 3.29.

Table 3.29: Awareness about Financial Incentive to construct toilet under Swachh Bharat Mission

Response	No. of Respondents	Percentage
Yes	253	36.4
No	423	60.9
Not responded	19	2.7
Total	**695**	**100**

Source: Field Survey, 2022-23.

The analysis showed that 36.4 per cent of the respondents had received subsidy for construction of individual household latrine under the Swachh Bharat Abhiyan. However, 60.9 per cent of the surveyed respondents stated that they did not receive the financial benefit under the mission. This could be due to the fact that those households were above poverty line and already had the latrine facility. Further, it was found that 2.7 per cent of the respondents did not prefer to respond on this aspect of the mission. The analysis shows that households are being benefited under the Swachh Bharat Abhiyan with financial support for construction of the latrine. Hoverver, there is a need to enhance the amount of subsidy to enable the poor households to construct people latrine facility as the unit cost of construction is high in the hills.

Conclusion

The analysis of the socio-economic profile of the surveyed households showed that most of the households were male headed and most of them belonged to the age group of 30-40 years followed by 40-50 years. Regarding the education level, it was found that 26 per cent of the surveyed households were illiterate and rest 74 per cent were literate. Primary occupation of most of the respondents was crop farming followed by business and government service. It was found that majority of the surveyed households had annual household income in the range of 5 lakh to 10 lakh and 2.5 lakh to 5 lakh. By religion, majority of the surveyed households were Christian followed by *Donyi Polo*. Tribe-wise distribution showed that most of the households belonged to *Nyishi* tribe. It was found that more than half of the surveyed households beloged to BPL category. Regarding family size, it was found that most of the households had semi-medium size of family (4-6 members) and majority of them had one earning member. Further, it was found that a high percentage of the surveyed households owned house and land indicating better economic security of tribal households of the district. At the same time, a good proportion of the surveyed households had other assets such as television, refrigerator, motor cycle and car.

The level of awareness of the surveyed households on various aspects of sanitation was measured by using a five-point Likert Scale. It was found that a good proportion (40-45%) of the respondents had good level of awareness on various aspects of sanitation and 25-30 per cent had very good level of awareness. Further, 5-10 per cent of the respondents had excellent level of awareness. But 15-20 per cent of the respondents had average level of awareness on various aspects of sanitation. This indicated that there is a scope for improvement in the level of awareness of the people on various aspects of sanitation. The mean score of awareness level of heads of the surveyed households on all the aspects of sanitation under consideration ranged between 3.16 to 3.31. This indicated that on an average the repondents had good level of awareness about the various aspects of sanitation. The mean score of awareness level was found to be the highest for effect of open defecation and the lowest for importance of drainage and waste segregation. It indicates that there is need to conduct awareness programme in different villages of the district to create awareness about the importance of sanitation in general and in particular, the need for having proper drainage and segregation of waste for achieving the goals of Swachh Bharat Mission. The analysis of awareness of surveyed households about the aims and provisions under the Swachh Bharat Abhiyan revealed that most of the respondents were aware about the mission. Further, it was found that a very high proportion of the respondents were of the opinion that everyone should participate in the Swachh Bharat Abhiyan activities. It was also found that more than one-third of the surveyed households had received financial incentive for construction of individual household latrine under the mission. However, the financial support received by the beneficiaries were not uniform. This calls for effective monitoring for proper implementation of the programme.

4
Households' Attitude towards Sanitation, Access to Sanitation Facilities and Hygiene Practices

Introduction

Access to sanitation is prerequisite for a healthy society and productive life of the people. It is essential not only for good health of the people but also for a healthy economy. According to a World Bank study, India is losing 4-6 per cent of Gross Domestic Product (GDP) due to damages to public health resulting from unsafe drinking water, lack of basic sanitation and environmental pollution (Nath, 2017). Hence, it calls for accelerating investment on water supply and sanitation in developing countries like India for improving public health and accelerating economic growth. The households' access to sanitation depends on a number of factors, such as income, occupation, education of members, attitude and awareness of the members as well as policy of the government. Attitude towards sanitation is one of the key determinants of access to sanitation facilities. A household with members (particularly household head) having positive attitude towards sanitation is likely to have better access to basic sanitation, irrespective of other factors. Therefore, the study has made an attempt to examine the attitude of the households towards sanitation. Further, the study also made an attempt to explore the extent of access of the households to basic sanitation such as safe water, individual household latrine, bathroom, drainage, dustbin and community or municipal bin. However, having access to sanitation facilities is not the sufficient condition for improving public health as it does not take into account the sanitary condition of the facility and its usage. The overall improvement in sanitation and public health, apart from access to sanitation facilities, depends on sanitary habits and hygiene practices of the people and the community (Nath, 2017).

Therefore, this chapter is devoted to analyse the households' attitude towards various aspects of sanitation, access of the households to sanitation facilities and sanitary habits and hygiene practices among the surveyed households in Papum Pare district based on primary data.

Households' Attitude Towards Sanitation and Hygiene

One of the prime concern of India is poor sanitation among the vast section of its society. Unsafe drinking water, poor hygienic conditions and improper disposal of human excreta are the big challenges which the country has been facing for many years. Sanitation and hygiene practices of a community depends on a number of factors. Among the various factors, these practices are greatly influenced by knowledge and attitude of the people towards sanitation (Mohd. and Malik, 2017). In fact, knowledge, attitude and practices (KAP) is an important way of knowing about hygiene behaviour and practices in the community. While knowledge refers to the information that people have about sanitation and hygiene, attitude is the way a person views or feels about the sanitation and hygiene related issues. Practices refers to the ways in which the people demonstrate their knowledge, skills and attitudes through their actions. Therefore, knowledge, attitude and practices are important for improving sanitation and hygiene among the people.

The attitude of the surveyed households towards sanitation and hygiene is discussed as follows:

Perception about Use of Boiled/Filtered Water

Safe drinking water is key to good health and stay free from diseases. Therefore, it is important to ensure that all the households have access to safe drinking water. But the access of a household to safe drinking water, to a great extent, is determined by the attitude and perception of the household members and its head. The analysis showed that most of the respondents had positive attitude towards the use of boiled and filtered water for drinking purpose in their households. The details are presented in Table 4.1.

Table 4.1: Perception of households about use of boiled/filtered water

Response/Attitude	No. of Respondents	Percentage
Yes	644	92.7
No	45	6.4
Don't Know	6	0.9
Total	**695**	**100**

Source: Field Survey, 2022-23.

The study found that 92.7 per cent of the household respondents had positive attitude towards the use of boiled and filtered water in their household which is quite encouraging. They perceived that every household should use boiled and filtered water in their households for a healthy life.

Perception about Covering of Water Containers

Water gets contaminated easily when it is exposed to dust particles and other agents. Hence, water needs to be kept safe in a covered containers. The analysis of the perception of the households showed that almost entire surveyed households had positive attitude towards it. The details are presented in Table 4.2.

It was found that 99.6 per cent of the household respondents perceived that water containers should be covered to keep water safe and stay healthy. This shows that there has been a significant change in attitude of the people towards sanitation in the State.

Table 4.2: Perception of households to keep water container covered

Response/Attitude	No. of Respondents	Percentage
Yes	692	99.6
No	1	0.1
Don't Know	2	0.3
Total	**695**	**100**

Source: Field Survey, 2022-23.

Perception Towards Toilet with Septic Tank

Proper disposal of human faeces is important to avoid spread of diseases. Hence, it is essential that all households should have a latrine with proper septic tank. However, many households have latrine without septic tank. The analysis of the attitude of the surveyed households towards toilet with septic tank showed that majority of them had positive attitude towards having a toilet with septic tank. The details are presented in Table 4.3.

Table 4.3: Perception of households towards toilet with septic tank

Response/Attitude	No. of Respondents	Percentage
Yes	684	98.4
No	9	1.3
Don't Know	2	0.3
Total	**695**	**100**

Source: Field Survey, 2022-23.

Table 4.3 shows that majority of the respondents (98.4%) perceived that households should have proper latrine with septic tank. This shows that surveyed households had positive attitude and perception towards sanitation facilities. It is likely to influence sanitary practices among the people and improve the sanitation of the villages.

Perception about Washing of Hands before Food and after Defecation

Most infectious germs enter human body through the contamination of hands. Hence, regular washing of hand before taking of food is essential to disinfect and stay safe from the diseases. The habit of washing of hands before food is influenced by attitude and perception towards sanitation. The analysis of perception of the surveyed households towards this aspect of sanitation showed that majority of the respondents had positive attitude towards it. The details are presented in Table 4.4.

Table 4.4: Perception of households about washing of hands with soap before food

Response/Attitude	No. of Respondents	Percentage
Yes	692	99.6
No	2	0.3
Don't Know	1	0.1
Total	**695**	**100**

Source: Field Survey, 2022-23.

Table 4.4 shows that most of the respondents (99.6%) believed that members of the households should wash their hands with soap before taking food and after the defecation. This implies that the households have right attitude towards this aspect of sanitation.

Perception about Proper Drainage Facility

Drainage facility in a household is essential to dispose household sewage. The lack of proper drainage can cause stagnation of sewage leading to breeding of mosquitos and germs which can cause malaria and other diseases. The analysis of the attitude of the households towards having a drainage facility showed that around three-fourth of the respondents believed that each household should have a proper drainage system for disposal of sewage.

Table 4.5: Perception of households about proper drainage facility

Response/Attitude	No. of Respondents	Percentage
Yes	530	76.3
No	156	22.4
Don't Know	9	1.3
Total	**695**	**100**

Source: Field Survey, 2022-23.

Table 4.5 shows that 76.3 per cent of the household respondents had positive attitude towards having a drainage facility. In other words, they perceived that each household should be equipped with proper drainage facility to dispose household's sewage and keep the surroundings clean. However, 22.4 per cent of the respondents believed that drainage facility is not much important for household sanitation and 1.3 per cent of them perceived that they are unsure about it.

Perception about Disposal of Waste in Municipal Bin

Household wastes are one of the major sources of environmental pollution. The household activities of production and consumption generate huge quantity of waste on a daily basis. Hence, it is of utmost important to develop a mechanism for collection as well as disposal of household wastes. The proper disposal of waste depends on the attitude of the people. So, an attempt was made to understand the attitude and perception of the people about disposal of waste in municipal bin. The analysis showed that most of the respondents had positive attitude towards waste disposal in community or municipal bin. The details are presented in Table 4.6.

Table 4.6: Perception of households about disposal of waste properly in municipal bin

Response/Attitude	No. of Respondents	Percentage
Yes	537	77.3
No	144	20.7
Don't Know	14	2
Total	**695**	**100**

Source: Field Survey, 2022-23.

Table 4.6 shows that 77.3 per cent of the household respondents had positive attitude towards disposal of waste in municipal bin. In other words, they perceived that each household should use the municipal or community bin facility to dispose household's waste and keep the surroundings clean. However, it was found that 20.7 per cent of the respondents believed that it is not much important for household sanitation to dispose waste in municipal bin.

Perception about Segregation of Waste

Environment has a limited capacity to assimilate wastes. The waste can be of different types—organic and inorganic. Organic waste is biodegradable and can be easily assimilated by the environment through the natural processes. However, there are certain wastes such as inorganic waste (polythene, chemical waste, electronic waste) which are hard to be assimilated by the environment. Hence, there is a need to segregate waste into organic and inorganic for better waste management.The analysis of households' attitude towards waste segregation showed that a high proportion of the respondents believed that waste should be segregated for better waste management.

Table 4.7: Perception about segregation of waste into organic and non-organic

Response/Attitude	No. of Respondents	Percentage
Yes	649	93.4
No	39	5.6
Don't Know	7	1
Total	**695**	**100**

Source: Field Survey, 2022-23.

Table 4.7 shows that 93.4 per cent of the household respondents had positive attitude towards segregation of waste into organic and inorganic. In other words, they perceived that each household should segregate waste before disposing it in the community bin for better waste management. However, it was found that 5.6 per cent of the respondents believed that it was not essential to segregate waste and 1 per cent viewed that they don't know about it.

Perception about Discouraging Use of Plastic Items

Plastic items and polythene bags are non-biodegradable. It takes millions of years for their decay and decomposition. Accumulation of plastic is one of the biggest challenges to environmental sanitation. The waste plastics choke up the drains and create barrier in disposal of sewage. Further, its accumulation on land makes the soil barren and unsuitable for cultivation. It also disfigures the beauty of the surroundings and reduces aesthetic value of a place. Therefore, there is a need to discourage the use of plastic items and polythene bags among the masses. Hence, an attempt was made to understand the perception of the surveyed households about the use of plastic items. The analysis showed that nearly three-fourths of the respondents perceived that there was a need to discourage the use of plastic items. The details are presented in Table 4.8.

Table 4.8: Perception about discouraging use of plastic items

Response/Attitude	No. of Respondents	Percentage
Yes	510	73.4
No	142	20.4
Don't Know	43	6.2
Total	**695**	**100.0**

Source: Field Survey, 2022-23.

Table 4.8 shows that 73.4 per cent of the household respondents had positive attitude towards discouraging use of plastic items in the society. In other words, they perceived that use of plastic items should be reduced by each household for improving environmental sanitation and have clean and green environment. However, it was found that 20.4 per cent of the respondents believed that it was not essential to discourage use of plastic items and 6.2 per cent of the respondents stated that they do not know about this aspect. This implies that the people need to be sensitised about the harmful effects of plastic on the environment.

Status of Households' Access to Sanitation Facilities

Households' access to sanitation facilities is prerequisite for a healthy and a happy society. Access to sanitation keeps the people healthy as well improves their productivity. It prevents the spread of diseases and helps to alleviate poverty. In view of enormous health, economic and social benefits of access to adequate sanitation facilities, it is essential to ensure that every household have access to basic sanitation facilities such as; individual household latrine, safe drinking water, regular water supply, proper drainage facility and sewage disposal system, proper facility for disposal of household waste. Hence, the study made an attempt to analyse the access of the surveyed households to basic sanitation facilities in the study area. The status of access of the households to sanitation facilities are discussed as follows:

Households' Access to Water Supply

Safe drinking water is the basic necessity of life. Water is also of utmost importance for maintaining sanitation. It is needed in day-to-day household activities of cooking, cleaning and washing. The supply of safe water is important for public health as the lack of safe water poses enormous health outcomes, globally 4 billion cases of diarrhoea, 2.2 million deaths per annum and 62.5 million disability-adjusted life years lost are attributed to lack of access to safe water (Nath, 2017). According to World Bank estimates 99.9 per cent of deaths in the world attributed to poor water supply occur in developing countries. Hence, there is a need to increase investment so as to provide safe water to all the households. At same time, there is a need to address the problem of unsustainability of water sources and their quality. In India, a large proportion of people did have access to safe drinking water. As per 2011 Census, only 75.71 per cent of the households in India had access to drinking water within a distance of 500 metres. In this background, an attempt has been made to examine the access to water supply among the households in the study area. The details are presented in Table 4.9.

Table 4.9: Distribution of households by access to water supply in the household

Response	No. of Respondents	Percentage
Yes	593	85.32
No	102	14.68
Total	**695**	**100**

Source: Field Survey, 2022-23.

The study found that 85.32 per cent of the households had access to water supply in their households. But 14.68 per cent of the respondents did not have access to water supply in their households. The analysis of main source of water among the households showed that most of the household had access to tap water followed by tube well/hand pump. The details of main source of water among the surveyed households is given in Table 4.10.

Table 4.10: Main source of water among the surveyed households

Source	No. of Respondents	Percentage
Tap water	299	43
Tube well/hand pump	115	16.6
Well	223	32
Pond	2	0.3
Lake/River	48	6.9
Spring	2	0.3
Any other	6	0.9
Total	**695**	**100**

Source: Field Survey, 2022-23.

Table 4.10 shows that 43 per cent of the households had tap water as the main source and 32 per cent of the respondents had well as the main source of water supply. Further, it was found that 16.5 per cent of the respondents had tube well/hand pump as the main source of water and 6.9 per cent of the households were relying on river and pond as the source of water. This implies that nearly

60 per cent of the households were relying on water source other than tap water for meeting their water requirement. It indicates that the supply of water by the government agency is not sufficient and so people have to explore alternative sources of water for their sustenance.

Households' Access to Water Filter

Water contains numerous germs and particles which poses risk to human health. Hence, water drawn from the nature needs to be filtered and treated to make it safe for human consumption. So, an attempt was made to examine the access of the surveyed households to water filter in their households. The analysis showed that a good proportion of the households had access to water filter in their households. The details are presented in Table 4.11.

Table 4.11: Access of households to water filter in their house

Response	No. of Respondents	Percentage
Yes	467	67.19
No	228	32.81
Total	**695**	**100**

Source: Field Survey, 2022-23.

Table 4.11 shows that 67.19 per cent of the households had water filter in their households. But it was found around one-third of the households (32.81%) did not have water filter in their households which is a matter of concern as it can severely affect the health of their family members. Hence, there is a need for appropriate policy intervention to encourage people to install water filter in their households.

Since a high proportion of the people did not have access to water filter in their household, the study attempted to unearth the reasons for not having this facility by the households. The details are presented in Table 4.12.

Table 4.12: Distribution of households by reasons for not having water filter

Response	No. of Respondents	Percentage
Lack of knowledge	8	3.51
Limited income	73	32.02
Distant market	3	1.32
Use bottledý/jar water	70	30.70
Any other	25	10.96
Not responded	49	21.49
Total	**228**	**100**

Source: Field Survey, 2022-23.

Table 4.12 shows that out of the total households which did not have water filter, 32 per cent of them stated that they lacked access to this facility due to limited income. Further, it was found that 30.7 per cent of them did not install water filter as they use packaged water. It was found that 3.51 per cent of the households did not have water filter due to lack of knowledge about its benefit and 10.96 per cent of them cited other reasons (mainly lack of awareness).

Households' Access to Latrine, Bathroom and Drainage Facilities

Access to basic sanitation facilities is the key to public health and stay free from diseases. One of the serious threat to human health is the contamination of water sources and surroundings due to practice of open defecation by the people. It is, therefore, of utmost important to ensure that each and every household has access to sanitary latrine with adequate supply of water, access of bathroom, water supply and proper drainage facilities. Access to sanitary latrine is quite limited in India. As per 2011 Census, 53 per cent of the households in the country did not have latrine facility. In Arunachal Pradesh, 38 per cent of the households in 2011 did have latrine. This indicates very poor level of sanitation in the State in 2011.

However, the sanitation coverage in the country has drastically improved under the Nirmal Bharat Abhiyan and Swachh Bharat Abhiyan. In this background, an attempt has been made to examine the access of the surveyed households to sanitation facilities. The details are presented in Table 4.13.

Table 4.13: Distribution of households by access to basic Sanitation Facilities

Facility	No. of Households	Percentage
Latrine with septic tank	634	91.2
Other Latrine (Pit)	46	6.6
No latrine	15	2.16
Bathroom Pucca	565	81.3
Bathroom Kutcha	77	11.1
Drainage	359	51.65

Source: Field Survey, 2022-23.

Table 4.13 shows that majority of the surveyed households had access to latrine with septic tank. It was found that 91.2 per cent of the households had access to sanitary latrine with septic tank and 6.6 per cent of the households had latrine without septic tank. This can be attributed to increase in awareness among the people and construction of latrines under Swachh Bharat Abhiyan. It is encouraging to note that only 2.16 per cent of the households had no latrine facility in their households. The analysis showed that around 10 per cent of lacked proper sanitary latrine. The findings of the study indicate that there has been drastic improvement in coverage of households with latrine in the last 12 years. It can be attributed to improvement in education, connectivity and various interventions by the government to bring about improvement in sanitation and quality of life of the people. But the quality of latrine sheds and their conditions were not up to the mark.

View of Latrines constructed under Swachh Bharat Abhiyan

The study found that 15 households (2.16%) did not have access to sanitary latrine in their households. Most of those households (66.67%) used shared latrine and some of them (20%) used community latrine. It was found that few of the surveyed households indicated other source of sanitary latrine like open space. Details are presented in Table 4.14.

Table 4.14: Type of toilet facility used by the households not having latrine

Facility used	No. of Households	Percentage
Community toilet	3	20.00
Shared toilet	10	66.67
Others	2	13.33
Total	**15**	**100**

Source: Field Survey, 2022-23.

Further, Table 4.13 shows that 92 per cent of the households had access to bathroom facility in their households. While 81.3 per cent of the households had access to *pucca* bathroom, 11 per cent of the households had *kutcha* bathroom.

Regarding access to drainage, it was found that 51.65 per cent of the surveyed households had drainage facility in their households (Table 4.13). It implies that around 50 per cent of the households lacked drainage facility in their households. Since drainage is an important component of sanitation for disposal of sewage, it is essential to encourage the households to build up proper drainage in their households. The analysis of type of drainage showed that most of the households had open pucca drainage. The details are presented in Table 4.15.

Table 4.15: Distribution of households by type of drainage in their households

Drainage type	No. of Households	Percentage
Underground	11	3.06
Covered *pucca*	32	8.91
Open *pucca*	224	62.40
Open *kutcha*	92	25.63
Total	**359**	**100**

Source: Field Survey, 2022-23.

Table 4.15 shows that out of 359 households having drainage, 62.4 per cent of them had access to open *pucca* drainage followed by open *kutcha* drainage (25.63%). It was found that only 8 per cent had access to covered drainage and 3 per cent of the households had underground drainage. The study shows that there is a much scope for improvement in drainage coverage in the study area for improving overall sanitation and quality of life of the people.

Source of Finance for Construction of Latrine

Construction of individual household latrine is essential for safe disposal of human faeces and prevent spread of diseases such as diarrhoea and typhoid. However, the construction of latrine involves fairly high cost. While well-off households can construct the latrine by financing from their own pocket, poor households with low income find it difficult to construct proper latrine. Therefore, in order to assist the poor households to construct the latrine, financial support (subsidy) is given by the government under the sanitation programme like Swachh Bharat Abhiyan.The analysis of the source of finance for construction of latrine showed that most of the surveyed households financed from their own pocket. The details are given in Table 4.16.

Table 4.16: Distribution of households by source of finance for construction of latrine

Source of Finance	No. of Households	Percentage
Own pocket	429	66.10
Loan from bank	1	0.15
Loan from relative/money lender	3	0.46
Govt. subsidy	9	1.39
Partly from own finance and subsidy	207	31.90
Total	**649**	**100**

Source: Field Survey, 2022-23.

Table 4.16 shows that 66 per cent of the surveyed households financed the construction of latrine out of their own pocket and 32 per cent of them financed partly from own finance and subsidy. Regarding the unit cost of construction of individual household latrine, it was found that the average cost of construction of latrine among the surveyed households was Rs. 1,42,000 and the subsidy amount under the Swachh Bharat Mission was Rs. 12,000. This implies that subsidy amount can finance only a small fraction (8%) of the total unit cost. Since the unit cost of construction of a proper latrine costs more than a lakh and subsidy amount is-meagre, it may be the reason for which the poor households are not able to build proper latrine facility. The cost of construction is more than two times higher in the hilly States as compared to the plain regions. Therefore, higher amount of subsidy may be provided to the poor people in the hills so that they are encouraged to build proper sanitary latrine in their households.

Households' Access to Waste Disposal Facility

Management of waste is an important requirement for improved sanitation in the community. Proper waste management is essential to improve public health as well as to maintain the beauty of the landscape of a village or a town. Therefore, there is a need for every village and town to have proper waste disposal facility as well as facilities for scientific treatment of waste. Lack of access to the waste disposal facility can compel the people to dispose waste in sensitive areas resulting in negative environmental externalities and human health hazards. So, it is very important that the government should provide each household with access to municipal or community waste disposal facilities. Hence, an attempt was made to examine the households' access to waste disposal facility. The analysis showed that only around one-third of the respondents had access to municipal or community waste disposal facility. The details are presented in Table 4.17.

Table 4.17: Distribution of households by access to community waste disposal facility

Response	No. of Households	Percentage
Yes	220	31.65
No	441	63.45
Not responded	34	4.90
Total	**695**	**100**

Source: Field Survey, 2022-23.

Table 4.17 shows that only 31.65 per cent of the surveyed households had access to municipal or community waste disposal facility. It was found that a very high proportion of the surveyed households (63.45%) did not have access to waste disposal facility. This could be one of the reasons for poor environmental sanitation and incidence of diseases in the rural areas of the State. It calls for attention of the policy makers to improve waste disposal facility in the rural areas and set up adequate number of waste treatment plants.

Households' Sanitary Habits and Hygiene Practices

Households' access to sanitation facilities is only a necessary condition for improved condition of public health. But mere having access to sanitation facilities is not the sufficient condition for achieving the goal of better public health. This is because access to sanitation facilities does not take into account the sanitary condition of the facility and its usage. For example, a household may have a toilet, but it is possible that it is not regularly cleaned or not kept in hygienic condition or it is not used due to certain reasons. Sengupta (2017) in his article '*Nirmal Grams: Problems and Prospects*' reported that most of the latrines provided to rural households under the Central Rural Sanitation Programme (CRSP) during 1980s were used by the villages for other purposes like storage for firewood and domestic animals etc. The study found that the main reasons for non-use of the latrines were—unfinished installation of latrines, lack of behavioural change, absence of super structure and blockage/chocked pan and lack of water. It is argued that the overall improvement in sanitation and public health, apart from access to sanitation facilities, depends on sanitary habits and hygiene practices of the people and the community (Nath, 2017). Hence, an attempt was made in this study to examine the sanitary habits and hygiene practices among the surveyed households.

The sanitary habits and hygiene practices among the surveyed households are discussed as follows:

Habit of Use of Toilet Cleaner by the Households

Regular use of toilet cleaner is an important sanitary and hygienic practice to stay safe and healthy. A household may have a toilet facility, but if it is not cleaned regularly using cleaners the health of its member may be at risk. Therefore, a household need to inculcate the habit and practice of use of toilet cleaner on routine basis to protect its members for various diseases. The analysis of the data showed that most of the respondents used toilet cleaner (Table 4.18).

Table 4.18: Distribution of households by use of toilet cleaner

Response	No. of Respondents	Percentage
Yes	639	91.9
No	56	8.1
Total	**695**	**100**

Source: Field Survey, 2022-23.

It was found that 91.9 per cent of the surveyed households used toilet cleaner in their households to keep the sanitary facility disinfected and safeguard the health of the family members. However, it was found that 8 per cent of the surveyed households were not using toilet cleaner. This implies that still there are some households whose members are at risk of being infected by various diseases despite having sanitary facility.

The analysis of the frequency of use of the toilet cleaner by the surveyed households revealed a very alarming picture. It was found that only 16.97 per cent of the surveyed households were using the toilet cleaner on a daily basis (Table 4.19).

Table 4.19: Frequency of use of toilet cleaner by the households

Frequency of use	No. of Respondents	Percentage
Daily	118	16.97
Alternate days	251	36.12
Once a week	270	38.85
Don't use	56	8.06
Total	**695**	**100.00**

Source: Field Survey, 2022-23.

Table 4.19 shows that 36.12 per cent of the surveyed households used the cleaner on alternate days and 38.85 per cent of them used once a week. This shows that there is a need for appropriate intervention to encourage people to use toilet cleaner on regular basis for improving the condition of sanitation and public health.

Habit of Use of Soap for Washing Hands by the Households

Soap and hand wash gel are an essential component of sanitary habit and hygienic practices. During the course of day-to-day activities, people touch a number of items and things through which there is a high risk of their hands getting infected by bacteria, germs and viruses. So, it is very important for all of us to wash our hands before certain important activities like before taking of food and eatables and after the defecation. During the period of COVID-19 pandemic in 2020 and 2021, the governments of all countries of the world issued directives to their people to frequently wash hands and use sanitiser as one of the weapons to fight against the Corona virus. In fact, the pandemic has helped in forcing the sanitary habit of washing hands and using sanitiser among the masses in developing countries like India and others. Since washing of hands with soap is considered as an important habit and hygienic practice to stay free from diseases, an attempt was made to examine this sanitary practices among the surveyed households. The study found that a very high proportion of the surveyed households used soap for washing hands before taking food. The details are presented in Table 4.20.

Table 4.20: Use of soap for washing hands by members of the households before food

Response	No. of Respondents	Percentage
Yes	588	84.60
No	41	5.90
Sometimes	66	9.50
Total	**695**	**100**

Source: Field Survey, 2022-23.

Table 4.20 shows that 84.60 per cent of the surveyed households used soap for washing hands before taking food and 9.50 per cent of them used it sometimes. It a matter of concern to note that 5.90 per cent of the surveyed households did not have the habit of using soap to wash hands before taking food. This calls for conducting an awareness programme on importance of washing hands.

Regarding the use of soap for washing hands after defecation, it was found that 90.65 per cent of the surveyed households used soap for washing hands after defecation (Table 4.21).

Table 4.21: Use of soap for washing hands by members of the households after defecation

Response	No. of Respondents	Percentage
Yes	630	90.65
No	27	3.88
Sometimes	38	5.47
Total	**695**	**100**

Source: Field Survey, 2022-23.

Table 4.21 shows that 5.47 per cent of the surveyed households used soap sometimes for washing hands after defecation. Further, it was found that 3.88 per cent of the households did not use soap for washing hands after defecation. It a matter of serious concern as the health of the members of those households are at risk. This calls for conducting an awareness programme to educate the people on the importance of washing handsafter defecation.

Habit of Segregation of Waste by the Households

The household sector generates huge quantities of solid waste. The habit of proper disposal and management of waste is important for keeping the household members safe from various diseases and enjoy a healthy life. The proper management of household waste require segregation of waste into organic and inorganic parts and regular disposal of the waste. The analysis of the waste segregation habit of the surveyed households revealed that most of the households practised the habit of segregation of waste. The details are presented in Table 4.22.

Table 4.22: Segregation of household waste into organic and non-organic

Response	No. of Respondents	Percentage
Yes	562	80.9
No	114	16.4
Sometimes	19	2.7
Total	**695**	**100**

Source: Field Survey, 2022-23.

Table 4.22 shows that 80.9 per cent of the surveyed households had the habit of segregation of waste into organic and inorganic parts. While 2.7 per cent of the households segregated waste sometimes, 16.4 per cent of them did not have the habit of segregation of waste.

Regarding the habit of keeping separate dustbin for organic and inorganic waste, it was found that majority of the surveyed households had separate dustbin (Table 4.23).

Table 4.23: Distribution of housholds by Separate dustbin for organic and non-organic waste

Response	No. of Respondents	Percentage
Yes	557	80.14
No	134	19.28
Sometimes	4	0.58
Total	**695**	**100**

Source: Field Survey, 2022-23.

Table 4.23 shows that 80.14 per cent of the surveyed households had the habit of keeping separate dustbin for organic and inorganic waste. But it was found that around one-fifth (19.28%) of the households did not separate dustbin for organic and inorganic waste. This calls for conducting an awareness programme to educate people on the importance of segregating waste and its proper disposal.

A household can dispose its waste in various ways. Earlier, there were hardly any community bin in the rural areas. So, people used to dispose waste by throwing in waterbodies and burning in the open field. Now-a-days, community bins have been built in rural areas to collect the household waste. The analysis of the data showed that more than half of the surveyed households used the method of burning in the field to dispose their household waste. The details are presented in Table 4.24.

Table 4.24: Distribution of households by ways of disposal of household waste

Ways of waste disposal	No. of Respondents	Percentage
Dispose in Municipalý/Community Bin	210	30.22
Dispose in riverý/pond	14	2.01
Dispose in open dumpsite	43	6.19
Bury in the fieldý/yard	360	51.80
Dispose in pit	22	3.17
Dispose in Municipal Truck	28	4.03
Others	18	2.58
Total	**695**	**100**

Source: Field Survey, 2022·23.

Table 4.24 shows that 51.80 per cent of the surveyed households had the habit of disposing waste by burning in the field. However, it was found that 30.22 per cent of the surveyed households had the habit of disposing waste in community bin and 4 per cent of them disposed in municipal truck. This shows the improvement in coverage of community bin in rural areas of the studied district. It is a matter of concern that 2 per cent of the households had the habit of disposing waste in the river and ponds.

Households' Feedback on Waste Management System

A waste management system is the process that organisations use to dispose of waste, reduce, reuse and prevent waste. It is an approach which an agency uses to efficiently manage waste from their origin until their final disposal. The various waste disposal methods could be composting, recycling, landfills, incineration, bioremediation, waste to energy and waste minimization. Thus, waste management is the measure adopted by the organisations or communities to manage waste from waste generation to final disposal or recovery. Environmental sanitation of an area requires a very effective, vibrant and scientific system of waste management. The proper management of waste is, therefore, essential to maintain cleanliness and hygiene of the community and safeguard the people against various diseases. It is also important to maintain the beauty of landscape of a place. Thus, it is important to ensure that every village and town has a very effective system of waste management. Hence, an attempt was made to assess the waste management system in the study area through the feedback of surveyed households. The details of the feedback of households on the waste management system of their areas are presented in Table 4.25.

Table 4.25: Households' feedback on waste management system of village

Response	No. of Respondents	Percentage
Poor	156	22.4
Average	251	36.2
Good	263	37.8
Very Good	12	1.7
Not responded	13	1.9
Total	**695**	**100**

Source: Field Survey, 2022-23.

Table 4.25 shows that majority of the respondents considered waste management system in their area to be average and poor. It was found that while 36.2 per cent of the households considered waste management system average, 22.4 of the households considered it to be poor. But it was found that 37.8 per cent of the households considered waste management system to be good. The analysis revealed that the waste management system in the villages study area needs to be improved for better environmental sanitation and public health.

Conclusion

On the basis of above findings and discussions, it can be concluded that improvement in public health is jointly determined by access to sanitation facilities and sanitary habits and hygienic practices among the people. The access to sanitation, on the other hand, depends greatly on the households' attitude towards sanitation. The study found that majority of the surveyed households had positive attitude towards various aspects of sanitation such as use of boiled and filtered water, covering of water containers, having toilet with septic tank, washing of hands with soap, having drainage in the household, disposal of waste in community bin and discouraging use of plastic items. The analysis of status of access to sanitation in the study area showed that majority of the households had access to water supply. However, it was found that 15 per cent of the households lacked access to water supply and so they were depending on natural sources of water like river, spring and pond. Regarding the source of water, nearly half of the households had access to tap water and a good proportion of them had well and tube well. While three-fourth of the households had water filter in their households, one-fourth of the households lacked this basic facility mainly due to limited income and dependence on packaged water.

The access to sanitary toilet showed that 97.6 per cent of the households had latrine facility and 92 per cent of them had bathroom facility. But 2.16 per cent of the households lacked latrine facility and most of those households used shared and community toilet. Few households which were not having latrine practiced open defecation. Further, it was found that more than half of the surveyed households had drainage facility (mostly open *pucca* type) in their households. Majority of the households had financed construction of latrine from their own pocket followed by partly from own pocket and partly by government subsidy. Access of the households to community waste disposal facility was found to be inadequate. This indicates that there is a lack of community bin in the rural areas. The sanitary habit and hygienic practices of the households revealed that majority of the households used toilet cleaner. However, the frequency of use of toilet cleaner was poor. It was found that most of the households had the habit of washing hands with soap before food and after defecation. But there was a scope for improvement in this habit. Regarding waste segregation it was found that fairly high proportion of households practised waste segregation into organic and inorganic. But there was a need for intervention to encourage households to inculcate the habit of waste segregation. The analysis of ways of waste disposal indicated that slightly more than half of the households disposed waste mainly by way of burning in the field followed by community bin. Some households stated that they dispose waste in the rivers or ponds. With regard to feedback on waste management system, most of the surveyed households considered waste management system in their locality to be average and poor.

5
Solid Waste Management System in Papum Pare District

Introduction

Waste and garbage management is one of the important components of environmental sanitation for improving the overall public health condition and protect the environment. Waste management has emerged as a big challenge for the public authority and the community in the last few years due to rapid increase in volume of the solid waste.The Solid Waste Management Rules (SWM) 2016 defines wastes as solid or semi-solid wastes consisting of domestic waste, sanitary waste, commercial waste, institutional waste, catering and market waste and other non-residential wastes, street sweepings, slit removed or collected from the surface drains, horticulture waste, agriculture and dairy waste, treated bio-medical waste excluding industrial waste, bio-medical waste and e-waste, battery waste, radio-active waste, which are generated in an area. The municipal wastes are broadly classified into three categories; biodegradable waste, recyclable waste and hazardous waste. The economic activities of production, consumption and exchange generate different types of waste in huge quantities. The primary reasons for increase in volume of solid waste and its accumulation in developing countries are: increase in population growth, expansion of economic activities and inefficient management of waste. As a result, the quantity of generation of solid waste is increasing rapidly. There is an urgent need for proper and scientific management of waste to protect human health and the environment.

India was the second most populous country in the world. Its population has been growing at a higher rate as compared to other emerging economies of the world. According to the United Nations estimates, India has surpassed China and became the world's most populous country with total population of 1.428 billion as on 1st April, 2023 (The United Nations, 2023). While China's population is estimated at 1.425 billion as on 1st April, 2023. The economy of India is also growing at a fairly higher rate and there has been significant expansion of economic activities in the country. All these factors have contributed to the generation of huge quantity of waste in the Indian economy. However, the investment in waste management in the country has been far below the mark and waste management system in the country has been far below the standard. India, currently, produces 277.10 million tonnes of solid waste every year, which is likely to reach 387.80 million tonnes in 2030 mainly due to population growth, rapid urbanization, mass consumption and economic development. In absolute term, India's solid waste generation is the highest in the world, but in per capita terms Bermuda and the United States top the list in the world. The per capita waste generation of India is 0.57 kg per person per day, which is below the global average of 0.74 kg per person per day and much below the per capita waste generation of Bermuda (4.54 kg per capita per day) and the US (2.24 kg per capita per day). However, the per capita waste generation in India is expected to rise to 0.90 kg per person per day by 2050, if appropriate steps are not taken to minimise waste generation (Santosh, 2021).

Among the cities of India, Delhi tops the list with waste generation of 30.60 lakh tonnes followed by Greater Mumbai (24.90 lakh tonnes) and Chennai (18.30 lakh tonnes) per year. The country produces 1.50 lakh tonnes of waste on daily basis, but only 20 per cent of the waste is processed and the remaining 80 per cent is dumped in the landfill sites in unscientific way (Santosh, 2021). The system of dumping of waste has led to creation of hills of waste in many cities of India like Delhi. This clearly shows inadequate investment in solid waste management and poor waste management system in the country. Thus, solid waste management will be a huge challenge to State governments and local municipal bodies in India in the 21st Century.

Arunachal Pradesh is a remote State of India. It is located in the easternmost corner of the country. It is one of the least populated States of the country with total population of only 13.83 lakh as per 2011 Census. It has the lowest population density, only 17 persons per sq. km (Census, 2011). In Arunachal Pradesh, majority of the people (77%) live in rural area and depend on shifting and other forms of cultivation, livestock rearing as well as forest based activities for their livelihood. Earlier, there was hardly any problem of solid waste management in the State. In the rural areas, most of the wastes used to consist of mainly organic biodegradable waste which were easily disposed-of in the field for producing compost. However, with the growth of urbanization, expansion of economic activities and consumerism the problem of solid waste management has emerged in both urban and rural areas of the State in recent years. Currently, there are more that 30 towns in the State. The largest towns, namely Itanagar and Naharlagun are located in the Papum Pare district. In fact, Papum Pare is the most urbanised district of the State with 55 per cent urban population. With modernisation and improvement in connectivity, the people in rural areas have also started to use the market goods and the plastic items. This has led to generation of different kinds of inorganic and non-biodegradable waste even in rural areas. The volume of waste generation has increased rapidly, but the waste management has not improved proportionately. The problem of waste management has emerged due to increase in volume of waste generation and inadequate efforts for waste management as well as lack of awareness and sanitary practices among the people.

In the last few years, several initiatives have been taken by the State government under the various schemes like Jawahar Lal Nehru National Urban Renewal Mission (JNNURM), Nirmal Bharat Abhiyan (NBA) and Swachh Bharat Abhiyan (SBA) for proper management of solid waste both in urban and rural areas. In this background, this chapter explores and evaluates the solid waste management system in Papum Pare district of Arunachal Pradesh.

Methods of Waste Management

Waste management is a collective activity involving segregation, collection, transportation, re-processing, recycling and disposing of different types of wastes. The different methods of waste management are:

Landfill: In this method, wastes are collected from around the cities which are then transported away from the city to the dumping sites and disposed-of. The dumping sites should be covered with soil to prevent contamination of water bodies. It also requires plantation of trees around the dumping sites to hold the soil and prevent spread of wastes. If designed carefully, it is one of the most economical and quite sanitised methods of waste management. But it is an unscientific method and unplanned dumping can cause serious environmental and health hazards like release of poisonous gases, secretion of toxic liquid and destruction of vegetations.

Incineration: In this method, wastes are burnt in the incineration plants at a high temperature enough to destroy the hazardous contaminants. It is useful to deal with large quantities of hazardous organic waste and produce electricity. But it is a costly method as the installation of incineration plant is expensive. It requires high cost equipment and skilled operators. At the same time, it generates ash and toxic gases (CO, SO_2 and HCL).

Pyrolysis: In this method, the solid waste is converted into liquid state and liquid is then converted into gas. These gases are then used for the production of energy.

Gasification: In this method, the waste materials to be treated are directly converted under high temperature into SynGas which has hydrogen and carbon dioxide as its components.

Bioremediation: In this method, living organisms like microorganisms are used to degrade environmental contaminants into less toxic form. For example, Pseudonymous bacterium can decompose synthetic pesticide. This technique is more cost-effective as compared to other methods as pollutants can be treated on-site, reducing exposure risk for the personnel.

Waste Management System in Urban Areas of Papum Pare

Papum Pare district is the most urbanised district of Arunachal Pradesh. As per 2011 Census, this district has a total population of 1.76 lakh, out of which 55 per cent consists of urban population which is far higher than the State average of 23 per cent. The twin capital towns of the State, namely Itanagar and Naharlagun are located in this district. These two towns are the main centres of education, healthcare, administration and marketing facilities. So, these two towns attract a large number of in-migrants for various purposes which has resulted in rapid growth of population and expansion of slums and colonies. The size and growth of population in the twin capital towns of Papum Pare district are presented in Table 5.1.

Table 5.1: Size and growth of population in twin Capital towns of Papum Pare district

Towns	Population (2001 Census)	Population (2011 Census)	Decadal growth
Itanagar	35022	59490	69.86
Naharlagun	27020	36158	33.82
Total	**62042**	**95648**	**54.17**

Source: Census of India, 2011.

The rapid growth of population and changes in consumption pattern of the people have been contributing to the generation of huge quantities of different kinds of waste in the urban areas of the district. Therefore, the solid waste management has emerged as a big challenge for the public authority in Itanagar and Naharlagun towns of Papum Pare.

Solid Waste Generation in Itanagar Capital Complex

Itanagar Capital Complex consists of three Circles, namely Itanagar, Naharlagun and Banderdewa and Borum Block. It is spread over an area of 200 sq. km (approx.). It has a total population of 1.23 lakh and a total of 105 villages (itanagar.nic.in, *accessed on 17the April, 2023*). Itanagar and Naharlagun are the twin capital towns of the State. The rapid growth of population has led to increase in number of households in Itanagar and Naharlagun. As per 2011 Census, there were 13,465 households in Itanagar and 7,735 households in Naharlagun. Household is one of the most important source of waste generation. Other sources of waste generation are: Institutions, Commercial establishment, Industries, Agriculture and natural sources. Among the various sources, the household or domestic source accounts for the largest share in waste generation. According to some estimates, domestic source contributes around 60 per cent of the total waste generation.

According to the report prepared by the Department of Urban Development and Housing, Government of Arunachal Pradesh, the garbage generation rate in Itanagar and Naharlagun was estimated at 0.450 kg per person per day (Tachung, 2021) and it produces around 50 MT of solid and wet garbage daily. Accordingly, the total waste generation per day in these two towns can be estimated by multiplying waste generated per person per day by the total population.

Table 5.2: Estimates of Waste Generation in Itanagar and Naharlagun

Towns	Waste generated per person per day (in kg)	Population (2011 Census)	Total waste generated per day (in MT)
Itanagar	0.450	59490	26.77
Naharlagun		36158	16.27
Total	**0.450**	**95648**	**43.04**

Source: Computed by Principal investigator using estimate of Tachung, 2021.

The total quantity of waste generation per day in Itanagar and Naharlagun based on the population as per 2011 Census was estimated at 26.77 MT and 16.27 MT respectively. The total waste generation per day in these two towns together was estimated at 43.04 MT. The analysis shows that huge quantities of waste are being generated per day in the twin capital towns. The above estimate is based on 2011 Census population and so the waste generation is substantially underestimated. Assuming the 2011 Census growth rate of population, the population of twin towns in 2023 is estimated at 1,57,627. Accordingly, the current waste generation in twin towns can be estimated as:

Total waste generation = 0.450 kg per person per day × 1,57,627 = 70.93 MT per day

Thus, the current waste generation in twin capital towns of Papum Pare district is found higher than the capacity of the treatment plant (50 MT).

Solid Waste Management in Itanagar Capital Complex: Past and Present Practices

Solid waste management was an unattended subject in Arunachal Pradesh until the year 2000. There was no system in place for management of solid waste in the State as a whole. The towns lacked facilities for collection, storage, transportation, processing and disposal of waste. There was no system of street sweepings. Further, there were no identified sanitary landfills and dumping grounds for these towns. However, since 2000, the system of solid waste management was started in the towns of the State. The State government started to give attention on the municipal solid waste management from the year 2000 onwards. All the district headquarters were provided manpower and machineries for waste management. A new Department of Urban Development and Housing (DUDH), created in April 1995 to formulate policies for urban development and creation of urban infrastructure, was assigned the task of solid waste management in towns of the State.

Currently, the task of solid waste management in Itanagar and Naharlagun is handled by the Itanagar Municipal Corporation (IMC).

The following systems of waste management were introduced in these towns:

(i) Street sweepings
(ii) Construction of dustbins at important locations for collection and storage of wastes.
(iii) Collection and lifting of waste from the collection points/ dustbins.
(iv) Transportation of waste to the dumping ground, but it was done in an unscientific ways.

However, the solid waste management was unscientific as the collection and lifting was irregular leading to stinking of unattended waste at the road sides and the dustbins of the towns. Due to lack of adequate dustbins in the capital towns, people used to throw their wastes on the side of the road and also in the rivers and canals flowing through the towns. The residents of Naharlagun town used to dump their waste in the *Lagun river* and some households were releasing the outlets of their toilets in that river. This had led to disfigurement of the towns and pollution of the rivers. Further, collected wastes were not segregated properly and dumped in the dumping sites in a haphazard manner. The dumping site for Itanagar and Naharlagun was located on the side of the Chimpu-Hollongi Highway which connects the State capital with Assam. The dumped municipal wastes were burnt in the open dumping sites which used to emit smokes and harmful gases in huge scale causing visibility problem and suffocation to the commuters due to pollution of air in the road side area. This system of waste management w s widely prevalent in Itanagar and Naharlagun towns for many years. It is noticed that now-a-days the practice of burning of waste in the dumping site has been discontinued as a result of opposition from the people.

View of Old Waste Dumping Site, Chimpu, Itanagar

Source: Tachung, 2021.

In the management of solid waste, special attention was given to the twin capital towns—Itanagar and Naharlagun. A project under Jawahar Lal Nehru National Urban Renewal Mission was sanctioned for setting up of a Waste Treatment Plant at Itanagar in 2007.

The details of Municipal Solid Waste Treatment Plant located at Chimpu-Hollongi road, at a distance around 10 km kilometre away from the Itanagar town is given below:

Table 5.3: Details of Municipal Solid Waste Treatment Plant, Itanagar

Capacity of the treatment plant	50 MT per day
Production of manure	15% of treated garbage (7.50 MT)
Source of Funding	JNNURM Scheme
Date of sanction of Project	23rd February, 2007
Date of start of the Project	5th April, 2009
Date of completion	7th March, 2013
Project area	13.61 acres
Built-up area	2 acres

Source: Tachung, 2021.

View of Municipal Solid Waste Treatment Plant, Chimpu, Itanagar in 2012

Source: Tachung, 2021.

View of Municipal Solid Waste Treatment Plant, Chimpu, Itanagar in 2023

Source: Photo taken by the Principal Investigator, Dr. Ashi Lama on 2nd April, 2023.

However, the waste treatment plant is lying defunct since 2014. It was started in 2013 and after functioning for some months, it stopped working due to technical and other reasons (Yigam, 2021).

So, the practice of dumping of municipal waste in the landfill or dumping zone located at Chimpu-Hollongi Highway in an un-engineered way is still being practiced in the twin capital towns. This practice is not safe for people travelling through this route and the residents. It can also pollute the environment and adversely affect human health in downstream and flora and fauna. The study team visited the site of waste treatment plant on 2nd April, 2023. The treatment plant was lying defunct and it was observed that the wastes were dumped in the vicinity of the plant and hilly slopes in a haphazard manner.

After the launch of *Swachh Bharat Abhiyan* (mission for cleanliness), the system of collection and storage of waste and garbage has changed in Itanagar and Naharlagun. The earlier system of collection of waste in dustbins and collection points has been discontinued in order to overcome the problem of accumulation of unattended waste and to maintain the cleanliness and beauty of the towns. The basic elements of new system of waste collection introduced in the twin capital towns are:

(i) Collection of waste from various colonies through the use of municipal dumpers. The households are required to put the waste in the morning hours directly in the municipal dumper instead of throwing in the dustbins.

(ii) To avoid the accumulation of wastes in the dustbins and the road sides, the dustbins were removed and throwing of waste on the road side collection points was strictly prohibited.

(iii) Door to door collection of waste by municipal staffs was also introduced in the two towns, but this system could not sustain due to lack of staffs.

(iv) Collection of fee from households for door to door waste collection service was also introduced to generate revenue. This system could not sustain due to lack of staffs with IMC as well as unwillingness of the households to pay the fee.

The new system of waste collection has helped to drastically reduce the problem of waste accumulation on the road sides and dustbins. Now, the people have developed the habit of carrying wastes up to the collection points of their respective colonies and directly throw their wastes on the municipal dumper. This implies that there has been significant improvement in sanitation and hygienic behaviour of the people and waste collection system in Itanagar and Nagarlagun. Now, the landscape of Itanagar and Naharlagun looks much better as compared to the past when the wastes were thrown in a haphazard manner along the road side of the towns.

However, there is no significant improvement in the system of waste disposal in Itanagar and Naharlagun. Since, the treatment plant is lying defunct, the waste collected are still being dumped in the landfill or dumping site in an unscientific manner. Thus, the solid waste management system in the twin towns as a whole is still unscientific. The twin capital towns are facing the crisis of waste management. Most of the solid waste is untreated and dumped in the open landfill. This happens because solid waste is dumped in low lying areas in the outskirts of the city. The public are not much aware about the system. Community bins are not designed properly, therefore, people throw waste anywhere at vacant places, poor facility of vehicle causes waste scattering during the time of collection and poor design of landfills causes the spread of pollution and negligence of public and lack of their cooperation makes this process more difficult.

The analysis shows that the municipal waste management system in the Itanagar capital region is still unsatisfactory and unscientific. The poor system of waste management had led to Itanagar being listed as one among the *top ten dirtiest cities* of India in the category of below 10 lakh population in 2020 (*The Arunachal Times*, August 21, 2020). So, the government directed Itanagar Municipal Corporation to take necessary steps to remove the tag (Eastern Sentinel, March 22, 2021). In view of growing population, urbanization and increase in generation of wastes in Itanagar and Naharlagun, there is an urgent need to revamp the treatment plant at Chimpu and make it operational. Further, there is a need to set up a separate waste treatment plant for Naharlagun town.

Recent Initiatives for Waste and Sewage Management in Itanagar

Taking cognizance of the poor management of waste in the twin capital towns, the National Green Tribunal (NGT) issued several important directions to the State Government regarding solid waste management in Itanagar Capital Region. The NGT asked that State Government to submit action taken report on the directives issued by it (Ajum, 2021). The Tribunal also suggested short-term measures to deal with solid waste. It suggested to set up adequate number of faecal sludge treatment plants which are easy to construct and also cost-effective.

The Itanagar Municipal Corporation (IMC) has proposed to set up two new solid waste management plants for twin capital towns, Naharlagun and Itanagar. The waste treatment plant for Naharlagun is being set up at Karsingsa. The site for the new solid waste management plant at Karsingsa was finalised in August 2021 (arunachal24.in). The NGT directed the IMC to complete the development of external infrastructure of landfill site at Chimpu by May 31, 2023.

In order to check the pollution of the rivers like Dikrong and Pachin, the IMC is setting up a 3 mld Sewage Treatment Plant (STP) funded by Government of India under the AMRUT 1.0 scheme for Naharlagun township. The NGT has directed the IMC to complete the project by September 2024. However, the progress of the construction of the plant is reported to be slow and it is suffering from delay.

Rivers are important sources of water. Therefore, for rejuvenation and cleaning of Senki and Pachin rivers, the IMC has submitted proposal to the government for Rs. 1,000 crore to set up sewage treatment plants.

In order to transport the sewage from different sources to sewage treatment plant (STP), IMC is laying the 'sewerage network pipes and sewerage pumps for pumping sewage water to reach the STP. The NGT has directed the IMC to complete the project by December 2023.

Waste Management System in Rural Areas of Papum Pare

The total population of Papum Pare district is 1.76 lakh, out of which 45 per cent live in rural areas (Census 2011). In this district, there were 485 villages and 35,730 Households as per the 2011 Census. Majority of the people practices shifting cultivation in the hilly slopes and settled paddy cultivation for their livelihood. In rural area, most of the waste is composed of organic which includes wastes from kitchens, gardens, cattle sheds and agricultural waste. In the villages, people have the traditional practice of segregating waste and using organic waste for composting. While the other wastes are either dumped or burnt in the open space. The waste management is looked after by the panchayats. With modernisation and integration of markets, villages are also witnessing the generation of different kinds of inorganic wastes. The waste management system in rural areas is still traditional. The most common practice is the burning of waste in the field and dumping in the forest and community dump site. However, it was found that in some villages located near the towns, wastes and garbage are collected by the municipal vehicles. The waste management practices in some of the surveyed villages of different Circles of the district are discussed as follows:

Waste Management System in Surveyed Villages of Itanagar and Naharlagun Circles

The existing system of waste management practices in the surveyed villages of Itanagar and Naharlagun Circles are discussed as follows:

Ganga Village

Ganga village is located at a distance of around 12 km away from Itanagar town. In this village, it was found that the waste collection vehicle comes to the village in the morning on alternate days basis with a catchy song of the Swachh Bharat Abhiyan *'Gadi wala aaya jara Ghar ka katchara nikaal'.* During the field visit, it was observed that the villagers were participating in the waste management by segregating most of their household waste, making compost with kitchen waste and adding it to the kitchen gardens. This practice is more common among the villagers. The waste collection vehicles have been provided on the efforts at the community level by gram panchayat. The village youth play a role in waste management by encouraging households to recycle organic waste. However, there is a need for encouraging more households to adopt composting and segregating non-organic through awareness programmes. The households' participation can be encouraged through awareness campaign and providing free distribution of segregating bins. It was stated that the village youth who look after the waste management system of the village with the support of gram panchayat face financial constraints as besides fuel they need to do the maintenance of the vehicle. So, the youth had decided to collect a fee of Rs. 100 per household per month to generate revenue as the fund provided by the State Government was not enough for waste management.

Chimpu Village

Many people from this village complained about the lack of awareness programmes for initiative of segregation of organic and non-organic waste. There is a lack of civic sense among the villagers about the garbage collection service being in place. Still garbage is dumped on the side of the main road. The respondents stated that people like to keep the inside of their home clean by pitching all their garbage on the public Street. The situation would not improve unless there is a change in behaviour and perception of the people about cleanliness. Every individual must contribute their part of job, only government is not responsible for the lapses. Another issue is the lack of proper drainage system. It was observed that there was an absence of adequate drainage system in almost every household of the village.

Kongko Village

In this village, the residents stated that funds were provided under various schemes for the development of sanitation system in the village. However, there is no visible improvement in cleanliness of the village. The residents stated that the unavailability of a garbage truck and proper drainage system to drain out sewage are the major problems. The lack of these facilities cause stinking of the locality and also promote breeding of mosquitos and germs. Secondly, water scarcity was another problem in the village. Safe water supply is the backbone of good health but its supply is not adequate. So, it needs to be prioritised. Afterall, the safe drinking water and proper sanitation is symbol of quality of life. Notably, the unavailability of proper road, electricity, drainage system and garbage truck are the core issues being faced by the residents this village.

Jollang Village

In this village, lots of accumulated wastes were noticed on the side of the street. The gram members of the village stated that the waste collection vehicle comes to colonies only once in a week. This is mainly due to the lack of adequate transportation facility, inadequate financial resources and inability to collect garbage fees from every household of the village, because most of the time they are dependent on State government for financial resources. They informed that besides the fuel, they also need to repair the vehicle involving high cost. So, they are not able to provide vehicle on daily basis for collection of waste. The members stated that they are trying to make the system more organised and better for collection and disposal of their garbage, but many villagers are refusing to give the garbage fee. This has affected the garbage collection effort. They stated that the unattended heap of garbage near the road side makes the village look ugly and also poses a huge health threat.

Sood Village

In this village, there is a provision of waste collection vehicle. But it is provided only once a week or on alternate day. So, the villagers sometimes dump their waste in the water bodies. However, there is practice of making compost with organic waste. The village lacks waste disposal facility.

Waste Management System in Surveyed Villages of Balijan Circle

The existing system of waste management practices in the surveyed villages of Balijan Circle are discussed as follows:

Nyishi Patila Village

This village is located near Hollongi. The waste management system in this village was found to be poor. The villagers stated they are yet to be provided with a garbage collection vehicle. This was due to fund issue. The villagers expressed their displeasure on non-provision of garbage collection vehicle. So, the wastes are dumped in different areas of the village like, forests and water bodies. Further, the drainage system in this village was found to be inadequate and poor. During rainy season, the village faces the problem of water logging causing lots of hardship to the villagers. The villagers were blaming the flawed construction of the road without proper drainage system. In some areas, the drainage system was constructed by the authority, but at many places the drains were found broken and blocked. The concerned authorities need to give immediate attention to address this problem as it can cause diseases like dengue, typhoid, jaundice. Almost all the villagers called for clearing the clogged drains.

Tubung Village

A lot of changes have taken place in waste management in this village. Earlier, the people used to throw garbage on the side of the village street. Now, people wait for the garbage pickup vehicle to throw their garbage. So, the garbage accumulation on the streets has reduced in the village. Earlier, villagers used to dump household garbage in the community dumping points, which was built by gram panchayat. Later, it was demolished due to insufficient equipment and lack of resources. They have started door to door garbage collection in the village. Cleanliness management is driven by collaboration of gram panchayat and self-help groups at the community level to sensitise households about the importance of segregation of wastes through awareness campaign. Now, most of the households segregate organic and inorganic waste. Waste collection movement is considered to be a game changer among the citizens under the Swachh Bharat Mission initiative. During the field survey, it was observed that the participants were doing their part by segregating most of their household waste. There were also some changes in the attitude and behaviour of all the villagers towards sanitation.

Upper Balijan

Upper Balijan village has a fairly well managed waste management system, which is looked after by the community. In the village, dustbins have been built in every 200 metres distance. The village panchayat has declared the village area as a plastic free zone. But there is a lack of community toilet and drainage system. Further, there is a lack of availability of proper water supply. Still some people directly depend on the rivers for drinking water. Streets are cleaned once a week by the members of self-help group. The village panchayat has strictly prohibited throwing of waste in the street and the rivers. Fines are imposed, if anyone is found to violate the rule.

Kokila Village

It is located near Hollongi. In this village, there was a lack of community dustbin and there was no provision of garbage collection vehicle. The access to water was also poor. Due to lack of waste disposal facilities, the households were found to dispose waste on the road side and the rivers. However, organic wastes from kitchen were used to make compost.

Waste Management System in Surveyed Villages of Sagalee Circle

The existing system of waste management practices in the surveyed villages of Sagalee Circle are discussed as follows:

Chumbang Village

The villagers usually dump their waste in the community dumping site. However, the practice of waste segregation was poor. The village did not have a proper drainage system. So, during rainfall grey water and garbage float along the village streets which causes diseases during the monsoon season. Earlier, unhealthy and unhygienic practices were common in the village. People did not wash their hands with soap after defecation. This transmitted various types of infections and diseases. Now, they have started using soap and hand wash for washing hand, but they need more education on the importance of segregation of organic and non-organic through awareness campaign. After Swachh Bharat Mission initiative, their lifestyle and behaviour has changed a bit, especially among the senior citizens. They have learned many things about sanitation. Now, they wash their hands with soap after defecation and keep their toilet clean.

Nyoriso Village

Collection of garbage is being done by the villagers themselves. The domestic waste is segregated at the household level itself by most of the villagers and organic waste is used in the agricultural field. The plastic and other non-biodegradable wastes are burnt down. But still there is no provision of garbage collection vehicles and the village lacks a comprehensive sanitation plan. Earlier, plastic and other wastes were thrown on the road sides. The lack of proper segregation of domestic waste into organic and inorganic in the household was due to lack of awareness. Hence, awareness is required way to improve the practice of waste management in the village. The segregation of waste does not require much manual labour and so it is cost-effective, but protects human health and the environment.

Yallang Village

In Yallang village, there is no system of door to door household garbage collection. But segregation of waste is done by the villagers themselves. There is a practice of using wet waste as compost and burning of the plastic and other wastes. They stated that there is a twin bin system to separate dry and wet waste. They appealed for provision of garbage carrying vehicles. Majority of the residents have been doing well on waste segregation and preparation of plans for effective sanitation management. The villagers stated that the waste management course should be taught in schools and colleges which would make the students understand about the importance of generating less waste in future. However, sometimes efforts do not sustain because of lack of cooperation and awareness on waste management procedures among the public.

Khamlee Village

Khamlee village has around 35-40 households and around 300 total population. It is situated at a distance of about 8 km away from Sagalee town. There was no proper waste management system. Plastic wastes were thrown on the roadside. There was a lack of proper maintenance of dustbin in most of the households and the shops. The village had a community bin but it was not used. There was no provision of garbage collection vehicle in the village. Further, the village had no drainage and sewage system. Community toilet was there but it was not used as there was no supply of water facility in the toilet. Hence, there is a need to conduct awareness campaign for waste management system and sanitation. Dustbin should be built in every 200 metres distance.

Gai Village

This village is located at a distance of around 12 km away from the block headquarter, Sagalee. In the village, the practice of segregation of domestic waste was not widely practised due to lack of awareness. There was no facility of waste collection vehicle and also no system of door to door collection of waste. The household wastes were disposed-of by way of burning and dumping on the natural sites. The village also suffered from the lack of access to safe drinking water. The villagers called for provision of dustbins, safe drinking water, a garbage collection vehicle, community dustbin, public toilet and conduct awareness campaign on sanitation.

Waste Management System in Surveyed Villages of Doimukh Circle

The existing system of waste management practices in the surveyed villages of Doimukh Circle are discussed as follows:

Rono Village

Rono is one of the villages of Doimukh circle. The village is located at a distance of only 2 km away from Doimukh town and thus it enjoys locational advantage. It has a fairly good maintenance of waste management system. The village is covered by community waste collection dumper everyday. Every houschold and business enterprise maintains personal dustbin. But there is a lack of community drainage system. The villagers urged for construction of a drainage system and suggested to build a dustbin in every 200 metres distance. They also demanded for regular provision of community garbage dumper in time.

Huto Village

Huto village is located only at a distance of around 5 km away from Doimukh town. The village is covered by the community waste collection dumper everyday. But there is a lack of community drainage system. The villagers urged for the construction of a drainage system and suggested to build a dustbin in every 200 metres distance. There is also a need for awareness campaign to encourage the people to adopt basic sanitation facilities rather than expect the government to do everything.

Waste Management System in Surveyed Villages of Banderdewa and Kimin Circles

The existing system of waste management practices in the surveyed villages of Banderdewa and Kimin Circles are discussed as follows:

Tarajuli Village, Banderdewa

In this village, the villagers were found to segregate household waste into organic and inorganic parts. The households use organic waste for making compost which is used in their agricultural field. Due to lack of dustbin and waste disposal facilities inorganic wastes were either thrown on the road sides or burnt in the open space.

Model Village, Kimin

In this village, there was community dustbins as well as community toilets. But in certain parts of the village those facilities were unavailable. It has a facility of garbage collection vehicle which comes every morning. It was observed that households use organic waste to make compost which is a good practice.

Durpa Village, Kimin

In this village, the facilities of community dustbins and community toilets were unavailable. Further, it had no facility of garbage collection vehicle. The villagers were anguished as despite repeated appeal, these facilities were not provided by the government. So, the households used traditional methods of waste disposal.

Key Issues in Solid Waste Management in Papum Pare

- Increase in waste generation due to rapid urbanization and mass consumption.
- Disposal of waste haphazardly in the dumping sites.
- Lack of good waste management practices, reduce, reuse and recycle (3Rs).
- Mixing of bio-medical waste with municipal waste.
- Lack of segregation of waste at the source.
- Inadequate attention to collection of construction debris.
- Lack of properly developed landfills and dumping sites.
- Lack of adequate and well-functioning waste treatment plants.

Challenges of Solid Waste Management in Papum Pare

- Lack of Awareness among the people.
- Rapid urbanization and population growth.
- Lack of proper drainage system to dispose-of sewage.
- Lack of Incineration plants and proper landfill.
- Disposal of wastes in the river and water bodies by the residents.
- Financial constraints—budgetary allocation is inadequate to properly manage the municipal and other waste.
- Inadequate Solid Waste Management Experts.
- Lack of Community Involvement.
- Stray Animals are also causing littering of the environment.
- Generation of huge quantities of plastic waste due to its extensive use as packing material and carry bags.
- Topographic and climatic constraints— the hilly topography and excessive rainfall for longer period during the monsoon season also pose problem in waste management.

Conclusion

The analysis and discussion indicates that the waste management system in both urban and rural areas of Papum Pare district is still inappropriate and unscientific. However, there has been significant improvement in waste collection system in Itanagar and Naharlagun towns. Before the year 2000, there was no waste management system in place in the State as a whole. The State government took certain steps to manage the waste in view of increase in volume of waste and its accumulation on the road sides. The government built municipal dustbins and waste collection points at various locations in the towns and dumpers were engaged to collect the garbage and waste. Recently, the government prohibited that throwing of waste on the road side and introduced a new system. Now, the people have to directly throw their waste in the municipal dumper. This has greatly reduced the accumulation of waste on the road sides.

However, there is no significant improvement in waste treatment and processing. The State initially adopted the landfill method of waste disposal that too in an un-engineered manner. Further, there was no system for waste segregation. The wastes collected from the towns were dumped at the dumping site located on Chimpu-Hollongi Road and those wastes were burnt in open site causing air pollution, visibility and suffocation problem for the commuters. An important step taken by the government was the setting up of a municipal waste treatment plant at Chimpu. The plant was started in 2013 for a few months only. But since 2004, it has been lying defunct and the wastes are disposed-of in the vicinity of the plants in a haphazard manner. The study found that the practice of waste collection has improved significantly, but there is no visible improvement in waste treatment. However, there is a ray of hope for the people as the two waste treatment plants, one at Chimpu-Hollongi Road and another at Karsingsa are being established by the Itanagar Municipal Corporation. In the rural parts of Papum Pare district, waste management system is improper and still traditional system of waste disposal, such as burning and dumping are being practised by the people due to lack of treatment plants and well developed landfills. However, the villagers have a good practice of composting of organic household wastes.

6
Summary and Conclusion

Introduction

This chapter presents the summary of the main findings of the study, highlights the key issues in sanitation. It also discusses the challenges of sanitation in Papum Pare district, in particular and Arunachal Pradesh in general, policy suggestions and conclusion. The identification of key issues and challenges of sanitation in the State are expected to help the policy makers in pin-pointing the areas of concerns and accordingly adopt appropriate interventions to address the key issues and overcome the challenges. The policy suggestions have been put forwarded on the basis of the main findings of the study. The suggestions are expected to be useful in formulating appropriate policy and adopting suitable strategies for improving access of the larger people and households to better sanitation facilities and develop sanitary practices among the people of the State and Papum Pare district.

The principal findings of the study are presented as follows:

Summary of the Findings

- It was found that as per 2011 Census, the percentage of households having access to individual household latrine in Arunachal Pradesh was 62 per cent which was higher than the national average of 46.9 per cent.

- However, in rural area of the State 47.3 per cent households were without latrine facility in 2011.
- Among the States of North East India, Mizoram had the highest percentage of households (92%) with latrine facility and Arunachal Pradesh had the lowest percentage of households (62%) with latrine facility in 2011 Census.
- The study found a high rural-urban disparity in access to sanitation facilities in the State. In rural Arunachal Pradesh, 47.3 per cent of the households were without latrine in 2011. But in urban area of the State only 10.5 per cent of households had no latrine.
- Regarding access to drinking water supply, it was found that as per 2011 Census only 65.5 per cent of the households in the State had access to tap water. In rural area, only 59.3 per cent of the households had access to tap water in 2011. Around 20 per cent of the rural households in the State were depending on river, spring, pond and lake for drinking water in 2011.
- Significant progress has been made under the Swachh Bharat Abhiyan. During the period 2014 to 2022, 10.9 crore toilets were built in the country. As per the report of National Annual Rural Survey 2019-20, 99.6 per cent of the rural households had toilet facility and availability of water—95.2 per cent households were using the facility.
- In Arunachal Pradesh, 1.45 lakh toilets were built under the Swachh Bharat Mission during 2014 to 2022.
- The analysis of coverage of the villages with drinking water revealed that in 2021, only 38.35 per cent of the villages in the State were covered with drinking water. While Dibang Valley had the highest percentage of villages covered with drinking water (66.42%), Kra Daadi had the lowest percentage of the villages covered with drinking water (14.45%).
- In Papum Pare district, it was found that in 2011 Census 81.15 per cent of the households had access to individual

household latrine. But the access to proper latrine was inadequate as only 43.93 per cent had access to latrine with septic tank.

- Further, a high rural-urban disparity was found in access to sanitation. In rural area of the district, only 58 per cent of the households had access to latrine. But in urban area, 96 per cent households had latrine facility.
- In rural area, access to proper sanitary latrine was low as only 27 per cent households had access to latrine with septic tank.
- Among the various Circles, Itanagar and Naharlagun had the highest percentage of the households having latrine facility. Kakoi, Mengio and Taraso Circles had the lowest percentage of the households having latrine facility. In Kakoi, only 4.58 per cent of the households were having latrine facility as per 2011 Census.
- Access to drinking water is important for well-being of the households. It was found that in Papum Pare district, only 56.91 per cent of households had access to tap water from treated source.
- High rural-urban disparity was found in access to water supply. In rural area of the district, the situation was worse as only 31.45 per of the households were having access to tap water from treated source in 2011 Census. But in urban area, 74 per cent of households had access to tap water from treated source.
- Among the various Circles, Itanagar and Naharlagun had the highest percentage of households with access to tap water from treated sources. But Parang and Kakoi Circles had the lowest percentage of households with access to tap water from treated sources.
- In *Parang Circle, only 0.35 per cent of the households* had access to tap water from treated sources and 70 per cent of the households were relying on river, spring, lake and pond for drinking water as per 2011 Census.

- The analysis of data on access to drainage in Papum Pare revealed that in 2011 Census only 61.66 per cent households has access to drainage facility. In rural area of the district, only 37 per households has access to drainage facility. But in urban area, a high proportion of the households (78%) households had access to drainage facility.
- Among the Circles, it was found that Naharlagun had highest percentage of households with drainage facility (76%) followed by Itanagar (73%). In Kakoi Circle, only 1.06 per cent of the total households had drainage facility in 2011 Census.
- Socio-economic profile of the surveyed households showed that most of the households were male headed and most of them belonged to the age group of 30-40 years followed by 40-50 years.
- Regarding the education level, it was found that 26 per cent of the surveyed respondents were illiterate and rest 74 per cent were literate. Primary occupation of most of the respondents was crop farming followed by business and government service.
- Majority of the surveyed households had their own house and land indicating better economic security of tribal households of the district. At the same time, a good proportion of the surveyed households had other assets such as television, refrigerator, motor cycle and car.
- Awareness level of the surveyed households on various aspects of sanitation was measured by using a five-point Likert Scale. It was found that a good proportion (40-45%) of the respondents had good level of awareness on various aspects of sanitation.
- It was found that 15-20 per cent of the respondents had only average level of awareness on various aspects of sanitation. This indicated that there is a scope for improvement in the level of awareness of the people on various aspects of sanitation.

- The mean score of awareness level was found to be the highest for effect of open defecation and the lowest for importance of drainage and waste segregation.
- It was found that most of the surveyed households were aware about the Swachh Bharat Abhiyan mission, its aims and provisions.
- Further, it was found that a very high proportion of the respondents opined that everyone should participate in the Swachh Bharat Abhiyan activities.
- It was also found that more than one-third of the surveyed households had received financial incentive for the construction of individual household latrine under the mission. However, the financial support received by the beneficiaries were not uniform. This calls for effective monitoring for proper implementation of the programme.
- The study found that majority of the surveyed households had positive attitude towards various aspects of sanitation, such as use of boiled and filtered water, covering of water containers, having toilet with septic tank, washing of hands with soap, having drainage in the household, disposal of waste in community bin and discouraging use of plastic items.
- The analysis of the status of access to drinking water in the study area showed that majority of the households had access to water supply.
- However, it was found that 15 per cent of the surveyed households lacked access to water supply and so they were depending on natural sources of water like river, spring and pond.
- Regarding the source of water, nearly half of the surveyed households had access to tap water and a good proportion of them had well and tube well.
- While three-fourths of the households had water filter in their households, one-fourth of the households lacked this basic facility mainly due to limited income and dependence on packaged water.

- The access to sanitary latrine facility showed that 97.6 per cent of the households had latrine facility and 91.2 per cent of the households had access to sanitary latrine with septic tank. This shows that there has been tremendous improvement not only in access to sanitary latrine but also access to proper latrine facility. This can be attributed to increase in awareness among the people and financial support for construction of latrines under Swachh Bharat Abhiyan.
- It was found that only 2.16 per cent of the surveyed households lacked latrine facility and most of those households used shared and community toilet. Few households which were not having latrine practised open defecation. Thus, Papum Pare is yet to achieve open defecation free (ODF) status.
- The study found that 92 per cent of the surveyed households had bathroom facility.
- Further, it was found that more than half of the surveyed households had drainage facility (mostly open *pucca* type) in their households.
- Majority of the households had financed the construction of latrine out of their own pocket.
- Access of the households to community waste disposal facility was found to be inadequate. This indicates that there is a lack of community bin in the rural areas.
- The sanitary habits and hygienic practices of the households revealed that majority of the households used toilet cleaner. However, the frequency of use of the toilet cleaner was poor.
- It was found that most of the households had the habit of washing hands with soap before food and after defecation. But there was a scope for improvement in this habit.
- Regarding waste segregation, it was found that fairly good proportion of households practised waste segregation into organic and inorganic. But there was a need for intervention to encourage households to inculcate the habit of waste segregation.

- The analysis of the ways of waste disposal indicated that slightly more than half of the surveyed households disposed waste mainly by way of burning in the field followed by community bin. Some households stated that they dispose waste in the river or pond. Majority of the respondents opined that the waste management system in their locality was average and poor.
- The study estimated that Itanagar Capital Complex currently generates around 71 MT of solid waste per day. But the municipal waste treatment plant is not functional.
- The analysis showed that the waste management system in both urban and rural areas of Papum Pare district is still inappropriate and unscientific. However, there has been significant improvement in waste collection system in Itanagar and Naharlagun towns in recent years.
- Since 2000 onwards, the State government took certain steps to manage the waste in view of increase in volume of waste and its accumulation on the road sides.
- Municipal dustbins and waste collection points at various locations of the twin capital towns Itanagar and Naharlagun were built and dumpers were engaged to collect the garbage and waste.
- Recently, the government prohibited throwing of waste on the road sides and introduced a new system. Now, the people have to throw directly their waste in the municipal dumper. This has greatly reduced the accumulation of waste on the road sides.
- However, there is no significant improvement in waste treatment and processing. The collected municipal waste are still unprocessed and dumped at the dumping site located at Chimpu-Hollongi Road as the municipal waste treatment plant at Chimpu is lying defunct since 2014.
- The study found that the practice of waste collection has improved significantly, but there is no improvement in waste disposal and treatment.

- The recent initiatives include two municipal waste treatment plants, one at Chimpu-Hollongi Road and another at Karsingsa which are being established by the Itanagar Municipal Corporation.
- In the rural areas of Papum Pare district, waste management system is improper and still traditional methods of waste disposal such as burning and dumping are being practised by the people due to lack of treatment plants and well developed landfill.

Key Issues in Sanitation in Papum Pare

- In adequate access of the rural households to proper sanitation facilities like individual households latrine, drinking water from treated source and drainage facility.
- High rural-urban disparity in access to sanitation facilities in Arunachal Pradesh as well as in Papum Pare district.
- Poor access of households to sanitation facilities like latrine and safe drinking water in the remote Circles of Papum Pare like Kakoi, Mengio, Parang and Taraso.
- Poor coverage of villages with supply of safe drinking water.
- Inadequate level of awareness among the households about various aspects of sanitation.
- Dependency of a good proportion of the surveyed households on rivers, springs and lakes for drinking water.
- Unhygienic and inferior quality of latrine facility in rural areas.
- Segregation of waste is still not up to the mark.
- Lack of municipal waste treatment plant and sewage treatment plant in Itanagar Capital Complex.

Challenges of Sanitation in Papum Pare

- Lack of awareness of the people about the ill effects of poor sanitation and improper waste management as well as inadequate sanitary habits.
- High incidence of poverty and low income of people in rural areas.

- Indifferent attitude of the people towards poor sanitation and waste management.
- Rapid urbanization, population growth and mass consumption due to online shopping leading to acceleration in generation of waste.
- Inadequate investment in provision of sanitation facilities and waste management.
- Delay in maintenance of existing plants and completion of new project for municipal waste treatment plants and sewage treatment plants.
- Lack of skilled and trained manpower in the field of sanitation and waste management.
- Tradition, customs and superstitions of the people also act as barriers to access better sanitation.

Policy Suggestions

The findings of the study lead to following policy suggestions:

- Organise awareness programmes on sanitation, hygiene and waste management in all villages and towns at regular interval to sensitise the people about the importance of sanitation and to change their mindset and behaviour towards sanitation and to make hygiene a cultural practice.
- Promote the principle of reduce, reuse and recycle (3Rs) among the masses and discourage the use of polythene and non-biodegradable items to mitigate waste generation.
- Focus on improving supply of safe drinking water and sanitary latrine in rural areas to reduce rural-urban disparity in access to water and sanitation.
- Special focus should be given on Circles of Papum Pare with poor access to drinking water and sanitation.
- Set up community waste treatment plants or scientifically developed Landfill in each panchayat for a group of villages.
- Introduce 'polluter pays' principle and impose pollution fee on each household and establishment to generate revenue. To make fee collection economical, the pollution fee may

be collected along with electricity or water bill and the amount may be transferred to the department taking care of sanitation and waste management.

- Adequate budgetary support should be given for sanitation and municipal solid waste management. Further, proper mechanism should be put in place for effective monitoring of the utilisation of fund.
- Introduce strong monitoring system to ensure timely completion of the ongoing projects on municipal solid waste treatment plant (SWTP) and sewage treatment plant (STP) in the Itanagar Capital Complex.
- Introduce courses on sanitation and waste management in Schools and Colleges.
- Establish training institute on sanitation and waste management to build up skilled and trained manpower.

Conclusion

Sanitation is indispensable for improving public health and ensuring healthy and productive life to the people at large. It enhances the quality of life as well as promotes social and economic well-being of the people. Access to proper sanitation facilities reduces healthcare cost, improves human productivity and thus, contributes to economic progress. It also enhances human dignity and ensures security and well-being of women and girl children. The overall improvement in sanitation and hygiene depends on access as well as on attitude, knowledge, sanitary habits and hygiene practices among the people. Sanitation is vital requirement for quality life, good health, safety, security, livelihood and economic development. But a huge proportion of people in developing countries like India are deprived of such basic facilities. The study found that despite efforts and significant improvement, access to safe drinking water and quality sanitation facility in the country is still below the mark. As per 2011 Census, more than half of the households in the country were without sanitary latrine and one-fourth of the households had no access to safe drinking water. In July 2021, 79 per cent of rural habitations were covered with drinking water. But there are issues of unsustainability of source and high operation and maintenance costs. In North East India, more than half of the habitations were uncovered with drinking water in July 2021.

In Arunachal Pradesh, sanitation has improved over the years, but there is a huge scope for its further improvement in all aspects of sanitation. Papum Pare district has a high sanitation coverage, but there was a high rural-urban disparity in access to safe water, proper latrine and drainage. Under Swachh Bharat Abhiyan, a significant progress has been made in terms of coverage with toilets. More than 90 per cent of the surveyed households had latrine facility. Most of the respondents were aware about the aims and provision of Swachh Bharat Abhiyan and this mission had helped to construct latrine in rural areas. The sanitary habits and hygiene practices among the Scheduled Tribes need to be improved through awareness campaign. The waste management system in the towns of Papum Pare district needs special attention as huge wastes are generated everyday. Though the waste collection system has improved in the Itanagar capital complex, yet waste disposal and processing is unscientific. In rural areas of the district, proper landfill or waste treatment plants should be developed to discourage dumping of waste in natural sites, roads and water bodies. The policy should focus on creating awareness and developing culture of hygiene among the people, special focus on rural areas, accelerate investment in water supply, operationalise waste treatment plants, impose pollution fee, disseminate knowledge on sanitation and build up skilled manpower to tackle the task of sanitation and waste management.

Annexure

Photos taken during the Field Survey in Different Villages of Papum Pare District, Arunachal Pradesh

Toilet Constructed under Swachh Bharat Abhiyan at Rono Village, Doimukh

Toilet Constructed under Swachh Bharat Abhiyan at Jully Village, Itanagar

Garbage Collection Vehicle at Ganga Village and Community Dustbin at Jollang, Itanagar

Garbage disposed on the Road Side at Amba, Doimukh

Water Supply at Upper Balijan Village

Public Toilet and Dustbin (SBA) at Upper Balijan

Toilet Constructed at Sagalee under SBA

Defunct Waste Treatment Plant and Dumping Site at Chimpu-Hollongi Road, Itanagar

Bibliography

Agarwala, T. (2015). Urban Sanitation in India: A Growth Story Gone Awry. *Yojana*, Vol. 59, pp. 28-32.

Ajum, B. (2021). NGT Issues Directions to GoAP regarding Waste Management. *The Arunachal Times*, September 30, 2021.

Akram, M. (2013). *Sanitation, Health and Development Deficit in India.* Paper Presented in the National Conference on Sociology of Sanitation-Environmental Sanitation, Public Health and Social Deprivation, organised by Sulabh International Centre for Action Sociology at Mavalankar Hall, New Delhi (January 28-29, 2013).

Akram, M. (2014). *Sociology of Health*. Rawat Publications, Jaipur.

Akram, M. (2014). Causes of Insanitation in India and Ways of Improvement. *Indian Plumbing Today*, November 2014.

Anbarci, N., Escaleras, M. & Register, Charles A. (2009). The III Effects of Public Sector Corruption in Water and Sanitation Sectors. *Land Economics*, Vol. 85, No. 2, pp. 363-377.

Basu, S.K. (1994). *A Health Profile of Tribal India*. https://pubmed.ncbi.nlm.nih.gov/12287763/

Chambers, R. & Medeazza, G.V. (2013). Sanitation and Stunting in India: Undernutrition's Blind Spot. *Economic & Political Weekly*. Vol XLVIII(25), pp. 15-17.

Chikarmane, P. (2015). Swachh Bharat Abhiyan: Marginalising the Mainstream, *Yojana*, Vol. 59, pp. 46-50.

Coffey, D., Gupta, A., Hathi, P., Spears, D., Srivastav, N., & Vyas, S. (2017). Understanding Open Defecation in Rural India: Untouchability, Pollution, and Latrine Pits. *Economic & Political Weekly*, Vol. LII (1). pp. 59-60.

Cronin, A.A. *et al.* (2015). *Gender Issues in Water and Sanitation Programmes, Lessons from India.* Sage Publications India, New Delhi.

Cullet, P. & Bhallar, L. (2015). *Sanitation Law and Policy in India.* Oxford University Press, New Delhi.

Franceys, R. & Gerlach, E. (2008). *Regulating Water and Sanitation for the Poor.* Earthscan, London, UK.

George, B. (2009). Sanitation Programmes: A Glass Half-Full, *Economic & Political Weekly*, Vol. 44(8). pp. 65-67.

Government of Arunachal Pradesh (2011). *Census of India 2011-Arunachal Pradesh District Census Handbook, Papum Pare.* Directorate of Census Operation, Itanagar.

Government of India (2011). *Census of India, 2011.* Ministry of Home Affairs, New Delhi.

Government of India (2022). *National Annual Rural Sanitation Survey 2019-2020,* Ministry of Drinking Water and Sanitation, New Delhi.

Government of India (2022). *Toilets Built Under Swachh Bharat Mission.* Press Information Bureau, Government of India, New Delhi, https://pib.gov.in/PressReleaseIframePage.aspx?PRID=1797158

Giri, R.R. (2016). Rural Cleanliness and Gandhiji. *Kurukshetra.* Vol. 64(12), pp. 39-42.

Gupta, V. & Pal, M. (2008). Community Sanitation Campaign: A Study in Haryana, *Economic & Political Weekly*, Vol. 43(33), pp. 20-23.

Heller, L. (2022). *The Human Rights to Water and Sanitation,* Cambridge University Press, UK.

Jana, S.K. & Bhowmick, P.K. (2003). Rural Sanitation Programme and Tribal Health: A Study in the District of Midnapore, West Bengal, *The Oriental Anthropologist*, Vol. 3 (1), pp. 74-77.

Jha, H. (2015). *Sanitation in India, A Historico-Sociological Survey*, Gyan Books Pvt. Ltd., Delhi.

Kumar, M. Dinesh, Bassi, N. & Kumar, S. (2022). *Drinking Water Security in Rural India, Dynamics, Influencing Factors, Improvement Strategy*. Springer Nature, Singapore.

Lasagasbaster, B. (2013).*Gender Perspectives on Sanitation for Sustainable Development*. https://www.devex.com/news/gender-perspectives-on-sanitation-for-sustainable-development-84889.

Mohan, R.V.R. (2017). Swachh Bharat Mission (Gramin): Bottlenecks and Remedies. *Economic and Political Weekly*, Vol. 42 (20), pp. 15-17.

Mohd., R. & Malik, I. (2017). Sanitation and Hygiene Knowledge, Attitude and Practices in Urban Setting of Bangalore: A Cross-sectional Study, *Journal of Community Medicine & Health Education*, Vol. 7 (4).

Nath, K.J. & Sharma, Vinod, P. (2017). *Water and Sanitation in the New Millennium*, Springer India Pvt. Ltd., pp. 1-8.

Nath, K.J. (2017). Overview of the current scenario of community water supply and environmental sanitation in India, In Nath, K.J. & Sharma, V.P. (eds.), *Water and Sanitation in the New Millennium*. Springer India Pvt. Ltd., pp. 1-8.

Pathak, B. (2016). Basic Sanitation in Rural India: Roadmap for Swachh Bharat. *Kurukshetra*, Vol. 64(12), pp .44-48.

Patel, A. (2016). Swachh Bharat Abhiyan to end Open Defecation in India. *Kurukshetra*, Vol. 64(12), pp .61-65.

Planning Commission (2013). *Poverty Estimation in India*. Planning Commission, Government of India, New Delhi.

Pradhan, G. (2015). Hygienic Practices among Tribal Communities: A Case Study of Odisha. *Tribal Studies*, Vol. 4 (2), pp. 236-238.

Ram, D. Sundar (2018). *Basic Sanitation in Rural India, Road Map for Swachh Bharat*. Shipra Publications, Delhi.

Rasure, K. A. (2019). Rural Drinking Water Supply and Sanitation. Manglam Publisher and Distributor, Delhi.

Reilly, K. (2010). Combining Sanitation and Women's Participation in Water Supply: An Example from Rajasthan. *Development in Practice*, Vol. 20(1), pp. 45-56.

Root, G.P.M. (2001). Sanitation, Community Environments and Childhood Diarrhoea in Rural Zimbabwe. *Journal of Health, Population and Nutrition*, Vol. 19(2), pp. 73-82.

Rout, A. (2016). Swachhta Pakhwada: When Government leads by example. *Kurukshetra*, Vol. 64(12), pp. 26-28.

Santosh (2021). Solid Waste Management: Types, Methods, Challenges and Solutions.https://www.youtube.com/watch?v=T_pIJiZ8JYI

Sengupta, C. (2017). Nirmal Grams: Problems and Prospects, In Nath, K.J. & Sharma, V.P. (eds.), *Water and Sanitation in the New Millennium*, Springer India Pvt. Ltd., pp. 21-40.

Shettar, S.C. (2013). Sociology of Sanitation: Incorporating Gender Issues in Sanitation, published in edited volume *Sociology of Sanitation*, Kalpaz Publication, New Delhi.

Singh, B. (2015). Sanitation, Development and Social Change: Cleaning of Holy Minds.*Yojana*, Vol. 59, pp. 70-73.

Singh, M.K. (2019). *Swachh Bharat Abhiyan*. Gyandeep Prakashan, New Delhi.

Srivastav, S. Saurabh (2022). *Sulabh Sanitation in India, Issues, Challenges and Prospects*. Rawat Publications, Jaipur.

Subbarao, V.P. & Somasekhar, S. (2015). Swachh Bharat: Some Issues and Concerns. *International Journal of Academic Research*, Vol. 2 (4), pp. 2348-7666.

Swain, P. & Pathela, S. (2016). Status of Sanitation and Hygiene Practices in the Context of Swachh Bharat Abhiyan in Two Districts of India. *International Journal of Community Medicine and Public Health*, Vol. 3 (11), pp. 3140-3146.

Tachung, G. (2021). *Solid Waste Management in Arunachal Pradesh—with Special Reference to the Solid Waste Management Project Implemented in the Capital Complex, Itanagar*. Department of Urban Development and Housing, Itanagar.

The Arunachal Times (2020). Itanagar among Top 10 Dirtiest Cities of the Country. *The Arunachal Times*, August 21, 2020.

The Eastern Sentinel (2021). Itanagar's waste management Issue to be taken up seriously: CM. *The Eastern Sentinel*, March 22, 2021.

Tiwari, R. & Nayak, S. (2017). Drinking Water, Sanitation and Waterborne Diseases. *Economic and Political Weekly*, Vol. LII(23), pp. 136-140.

Tomar, N. Singh (2016). Swachh Bharat Mission from Behavioural Change to Social Change. *Kurukshetra*, Vol. 64(12), pp. 5-7.

UNDP (2022).*Global Multidimensional Poverty Report (MPI)*. United Nations Development Programme.

United Nations (2023). *Revision of the World Population Prospects 2022*. The United Nations, Geneva.

Yigam, J. (2021). Solid Waste Treatment Plant at Chimpu Needs Early Restoration. *The Arunachal Times*, September, 9, 2021.

Websites and Links

https://jalshakti-ddws.gov.in

https://www.ncbi.nlm.nih.gov/pubmed/1228776

http://citeseerx.ist.psu.edu/viewdoc/download?doi=10.1.1.684.727&rep=rep1&type=pdf

https://www.researchgate.net/publication/215463031_Rural_Sanitation_Programme_

Index